Environmental Footprints and Eco-design of Products and Processes

Series Editor

Subramanian Senthilkannan Muthu, Head of Sustainability—SgT Group and API, Hong Kong, Kowloon, Hong Kong

Indexed by Scopus

This series aims to broadly cover all the aspects related to environmental assessment of products, development of environmental and ecological indicators and eco-design of various products and processes. Below are the areas fall under the aims and scope of this series, but not limited to: Environmental Life Cycle Assessment; Social Life Cycle Assessment; Organizational and Product Carbon Footprints; Ecological, Energy and Water Footprints; Life cycle costing; Environmental and sustainable indicators; Environmental impact assessment methods and tools; Eco-design (sustainable design) aspects and tools; Biodegradation studies; Recycling; Solid waste management; Environmental and social audits; Green Purchasing and tools; Product environmental footprints; Environmental management standards and regulations; Eco-labels; Green Claims and green washing; Assessment of sustainability aspects.

K E K Vimal · Sonu Rajak · Vikas Kumar ·
Rahul S. Mor · Almoayied Assayed
Editors

Industry 4.0 Technologies: Sustainable Manufacturing Supply Chains

Volume 1—Theory, Challenges, and Opportunity

Springer

Editors
K E K Vimal
Department of Production Engineering
National Institute of Technology
Tiruchirappalli
Tiruchirappalli, India

Vikas Kumar
Faculty of Business,
Law and Social Sciences
Birmingham City University
Birmingham, UK

Almoayied Assayed
Water, Environment and Climate
Change Centre
Royal Scientific Society
Amman, Jordan

Sonu Rajak
Department of Mechanical Engineering
National Institute of Technology Patna
Patna, India

Rahul S. Mor
Business Systems and Operations
Faculty of Business and Law
University of Northampton
Northampton, UK

ISSN 2345-7651 ISSN 2345-766X (electronic)
Environmental Footprints and Eco-design of Products and Processes
ISBN 978-981-99-4821-5 ISBN 978-981-99-4819-2 (eBook)
https://doi.org/10.1007/978-981-99-4819-2

This Springer imprint is published by the registered company Springer Nature Singapore Pte Ltd.
The registered company address is: 152 Beach Road, #21-01/04 Gateway East, Singapore 189721, Singapore

Foreword

Climate change issues are the major challenges of today's mankind, the root causes of which include burning too much fossil fuel for energy production, producing enormous food waste and dumping it in landfill, and contaminating our water sources. A circular economy is the paradigm shift for transforming a linear supply chain (make, use, and throw) to a circular (take, make, distribution, use, and recover) giving equal emphasis to economic, environmental, and social performance using reduce, reuse and recycle principle, adopting resource and energy efficiency, renewable energy and waste management across the supply chain. Technology (both conversion and communication) is one of the enablers for transforming the supply chain. Although larger organizations have effectively implemented a circular economy, measures for carbon footprint reduction in their supply chains within small and medium-sized enterprises are lacking. This book addresses these issues and focuses on the transformational aspects of industrial supply chains using communication technologies (e.g., Industry 4.0, Blockchain, IoT, and other tools).

The primary focus of the book revolves around sustainable design, planning, implementation, and operations of the industrial supply chain. It extensively covers various aspects such as challenges, opportunities, and issues while utilizing Industry 4.0. The book encompasses fundamental principles of Industry 4.0 and explores the practical implementation of AI, Big Data, Procurement 4.0, Logistics 4.0, Lean 4.0, and Machine Learning to foster resilience within the supply chain. It not only provides the theoretical foundation for sustainable supply chain operations but also delves into the practical implications of adopting Industry 4.0. Furthermore, the book sheds light on the current state of Industry 4.0 adoption in the industrial supply chain, highlighting associated challenges, opportunities, and factors that enable improvement. The transformation of the supply chain through Industry 4.0 necessitates cultural, technological, organizational, and structural modifications. Additionally, the book identifies emerging research questions that require attention in order to further promote the adoption of Industry 4.0 in the sustainable manufacturing supply chain.

The authors and editors possess extensive expertise in sustainable supply chains and the implementation of Industry 4.0. This book delves into case studies from both developed and emerging economies within the manufacturing industry. On the one hand, it presents theories on operations and supply chain management, while on the other, it showcases methods, frameworks, and practical applications of tools and techniques for organizational transformation in adopting Industry 4.0 to address climate change challenges. Having collaborated with the editorial team on numerous sustainability-related projects, I have witnessed their exceptional competence in research design, planning, and execution. The inclusion of real-world cases throughout the chapters enhances comprehension of the topic's theoretical foundation and its practical implications. I firmly believe that this book will captivate students, academics, and professionals alike. Thus, I extend my heartfelt congratulations to all those involved in bringing this project to fruition.

Prasanta Kumar Dey
Professor of Operations Management
Aston Business School
Aston University Birmingham, UK

Preface

Industry 4.0 technologies, representing the fourth industrial revolution, help businesses be more transparent and competitive, balancing resilience and sustainability in manufacturing supply chains. Though the literature comprehends some theoretical frameworks, challenges and opportunities for sustainable manufacturing supply chains, further insights are required. Therefore, this book presents various aspects of Industry 4.0 for the manufacturing supply chains, and includes contributions focused on adopting such technologies, challenges, opportunities, implementation frameworks and requirements to create a sustainable and resilient manufacturing business environment.

Chapter 1 outlines the digital supply chain towards technological advancement and enablers. Chapter 2 investigates the applications of digital technologies like artificial intelligence and data analytics and the critical challenges for managing supply chain decision-making. A pilot study to demonstrate the use of blockchain technology for tracking and ensuring the compliance of olive oil in Jordan has been documented in Chap. 3. A significant improvement towards sales, transparency and export capabilities is gained through this study. Chapter 4 suggests connecting sustainable manufacturing and Logistic 4.0 towards sustainable value creation and a triple-bottom-line viewpoint. Chapter 5 offers a case study from the food sector in Jordan focused on assessing Industry 4.0 technologies in resource-efficient and cleaner production practices and helps to monitor carbon footprint accurately. A holistic understanding of recent advancements and applications of machine learning and how these techniques can enhance supply chain operations is presented in Chap. 6. Chapter 7 highlights the benefits of using the fourth industrial revolution framework of technologies and their enabling components in the manufacturing sector using a mixed methods approach. Chapter 8 presents an overview of the digital twin to build an efficient supply chain along with its benefits and challenges in adopting the digital twin.

Chapter 9 provides a critical discussion and SWOT analysis of Industry 4.0 technologies like big data, Internet of things, additive manufacturing, and blockchain technology for the sustainable supply chain management. On the other hand, Chap. 10 analyses the modern theories, available information, and research gaps through a bibliometric study. Insights on the performance metrics to supply chain management activities are provided in Chap. 11. Various key performance indicators and their significance towards the digital supply chain have been elaborated. Finally, Chap. 12 highlights many general and technology-specific hurdles of implementing Industry 4.0 technologies and emphasizes addressing these challenges to enable the effective incorporation of Industry 4.0 technologies for sustainability goals.

Tiruchirappalli, India — K E K Vimal
Patna, India — Sonu Rajak
Birmingham, UK — Vikas Kumar
Northampton, UK — Rahul S. Mor
Amman, Jordan — Almoayied Assayed

Acknowledgement

Writing a book necessitates a substantial time commitment above all else, often leading to the sacrifice of personal time that could have been allocated to other responsibilities. As a result, we wholeheartedly express our heartfelt gratitude to our family members, whose unwavering support and encouragement have been invaluable throughout this endeavor. Additionally, we extend our sincere appreciation to our mentors and advisors, whose guidance and backing over the past few years have played a pivotal role in shaping us into the individuals we are today. We would also like to acknowledge and thank our home institutions for their support in undertaking this project. Lastly, we wish to express our gratitude to the Royal Academy of Engineering and the Newton-Khalidi Fund Transforming Systems through Partnership (TSP) program for partially facilitating this collaborative effort.

Contents

Digital Supply Chain Paradigm 1
M. Nishal, K. Ram Prasad, and R. Kumanan

Supply Chain Decision-Making Using Artificial Intelligence and Data Analytics 25
Lukman A. Akanbi, Kayode I. Adenuga, and Hakeem Owolabi

Using Blockchain for Agro-Food Traceability: A Case Study from Olive Oil Industry 35
Abdlerahman Habashneh, Almoayied Assayed, and Amro AlMajali

Logistics 4.0 for Sustainable Manufacturing Supply Chain 47
L. Aravindh Kumaran, M. Ramasubramaniam, and K. Sivakumar

Industry 4.0 in Resource Efficient and Cleaner Production: A Case Study from the Food Sector in Jordan 61
Almoayied Assayed, Jehan Haddad, Husam Kilani, Rawia Abdallah, and Vikas Kumar

Applications of Machine Learning in Supply Chain Management—A Review 73
P. Thejasree, N. Manikandan, K E K Vimal, K. Sivakumar, and P. C. Krishnamachary

The Benefits of Using Industry 4.0 in the Manufacturing Sector 83
Asad Ullah, Shahid Imran, and Deena Roy

Application of Digital Twin for Efficient Supply Chain: Analysis of Opportunities and Challenges 105
Laxmi Pandit Vishwakarma and Rajesh Kr Singh

Industry 4.0 Technologies: Opportunities in the Sustainable Supply Chain Management 125
Deepak Datta Nirmal and K. Nageswara Reddy

A Bibliometric Analysis of Smart Manufacturing and Way Forward ... 137
Saurabh Tiwari and Shantanu Trivedi

Performance Metrics in Digital Supply Chain Paradigm 159
R. Prashanna Rangan, R. Kumanan, K. Ram Prasad, and M. Nishal

Challenges for the Adoption of Industry 4.0 in the Sustainable Manufacturing Supply Chain 175
Hakeem Owolabi, Luicija Juryte, and Lukman Akanbi

Digital Supply Chain Paradigm

M. Nishal, K. Ram Prasad, and R. Kumanan

Abstract Technological advancement and globalization have created the opportunity to compete and flourish in business. For organizations to maintain sustainability and competitiveness at the global level in the corporation of digital technologies has become inevitable and the fourth industrial revolution has changed the way of communication and interact with their environment. The operational processes involved in the production and delivery of goods and services to clients are taken care of independently at the current organizational levels. The orthodox supply chain has facilities in different geographical locations to assist in establishing and maintaining logistics transportation connections within them. Supply chain management is always evolving as a result of the quickening development and changes in many markets and the economic, financial, social, and technological spheres. The supply chain will never be static but continuous changes in shape, size, configuration, and the way it is coordinated, controlled, and managed but never been static. The impact of the industrial revolution and supply chain applications was accelerated that made a transformation that is essential for the digital conversion of the organizations. In this chapter, the key constituents of supply chain management are described and through extended works of literature, the digital technology enablers are identified which would enhance the supply chain constituents. An insight into value drivers with key digitization technology for enhancing the supply chain management is discussed in detail which would enable supply chain analysts and researchers to have a deeper analysis of them in the context of the digitization of SCM.

Keywords Supply chain management · Digitalization · Digital supply chain · Industry 4.0 · Value chain drivers

---

M. Nishal (✉) · K. Ram Prasad · R. Kumanan
Department of Mechanical, Sri Venkateswara College of Engineering, Chennai, Tamilnadu, India
e-mail: nishal@svce.ac.in

K. Ram Prasad
e-mail: ramprasadk@svce.ac.in

K E K Vimal et al. (eds.), *Industry 4.0 Technologies: Sustainable Manufacturing Supply Chains*, Environmental Footprints and Eco-design of Products and Processes,
https://doi.org/10.1007/978-981-99-4819-2_1

1 Introduction

The introduction and wider usage of consumer gadgets like mobiles, computers, drones, wearables, and tablets have completely changed the way of information sharing. Be it using voice-controlled intelligence to real-time tracking of consumer goods, technologies used through these gadgets have minimized labor costs by streamlining data capture and reducing error-prone manual processes. With the availability of low-cost sensors and computing power, companies are able to generate, store, and analyze large volumes of data. These data provide insight into business management to take necessary action to mitigate loss and generate revenue [11].

The supply chain has started to be digitalized to embrace digital products and services as the process within the companies is undergoing these rapid changes. In the context of gaining an advantage in the digital supply chain, novel approaches are necessary including technologies for digital transformation. From the economic, environmental, and social perspective, the present global supply chain management (SCM) and logistics policies of freightage, stowage, and managing cargo are not sustainable for the upcoming scenario [20]. Supply chain and logistics operations have become a critical part of a day-to-day routine for administrative and individual activities and they contribute to global development. A conventional supply chain is a sequence of interconnected connections involving collaboration, outlining, control, and services among the customer and supplier. Supply chains are usually multitiered and sometimes involve multiple geographies requiring specialized raw materials and components, which depend on other services for effective functioning. A report from Mckinsey states that a conventional supply chain creates greater social and environmental costs than its own operations. More than 80% of greenhouse gas emissions and 90% of the impact on air, land, water, biodiversity, and geological resources are due to the supply chain activities performed related to the consumer's company, which has raised questions about the self-sustainable nature of the conventional supply chain [31]. Many technology giants have moved to digital platforms for their supply chain activities. For example, Google's digital twin uses real-time data received from sensors and the Internet of Things (IoT) to develop a digital replica of the physical supply chain. This can create dashboards and alert uses to critical users like potential disruptions [32]. On the other hand, Amazon which is already an online retailing business has integrated millions of buyers and sellers through its AWS (Amazon Web Services) cloud infrastructure to build a cloud-based data warehousing for managing its logistics operation [33]. Other companies which are in existence for decades and more are forced to change to incorporate digital technologies for competition in the current era. The term digitization is creating the digital version of paper documents, images, sounds, values, and more. That is converting the non-digital component to a digital layout to be used with calculating systems for numerous activities. This would result in an uninterrupted flood of information in digital form which directs to fast transfer and operation authenticity in turn creating a baseline for digitalization and automation of the operation process [2, 5].

The blockchain has revolutionized the supply chain by providing an efficient and transparent way to track the product in the supply chain. Blockchain is also used for ensuring product authenticity and origin. This digital ledger technology has enabled different categories of business for good. The implementation of QR codes with blockchain technologies has enabled cheap solutions to mitigate counterfeit medicine in African countries [18]. Because of the availability of huge reliable industry data, reduction of cost in computational devices, reliable internet connectivity (CPS), development of usable machine learning algorithms, the popularity of data science concepts to analyze data, and willingness of top management to adopt results of an analysis to improve business has been a paradigm shift from conventional human expert depended on methodologies [28].

2 What Supply Chain Management Is All About?

Supply chain management encloses all the methodologies that metamorphose the supplies into manufactured goods by administering the movement of goods and services. SCM demands the conscious universalization of a company's supply-side administration for enhancing customer services and coming into possession of a gladiatorial boundary in the market.

The group of people, undertakings, assets, challenges, and progression of technology utilized to forge and sell goods or services is known as a supply chain [1]. When the supplies are handed over from a seller to an industrialist and the ultimate good or service is shipped to the consumer is where the supply chain begins.

Supply chain management is a critical process because it can help in earning many professional missions. Managing production methodology, for example, we can increase the commodities attribute by reducing the probability of remembering and legitimate action and supporting the development of a powerful consumer brand. Control over shipping processes can also enhance customer service by preventing expensive shortages or periods of inventory overproduction [1].

Implementation of SCM varies for each company as each company's SCM process is unique due to its defined objectives, limitations, and advantages. Figure 1 shows the overview of various phases of supply chain management.

2.1 Upgradation to Digital Supply Chain Management

Digital supply chain provides value to its clients through the adoption of new technologies to gradually alter how the company runs and provides value to its clients. Digital supply chains have a wide range of advantages for businesses, including:

- Improved effectiveness;
- Lower expenses;

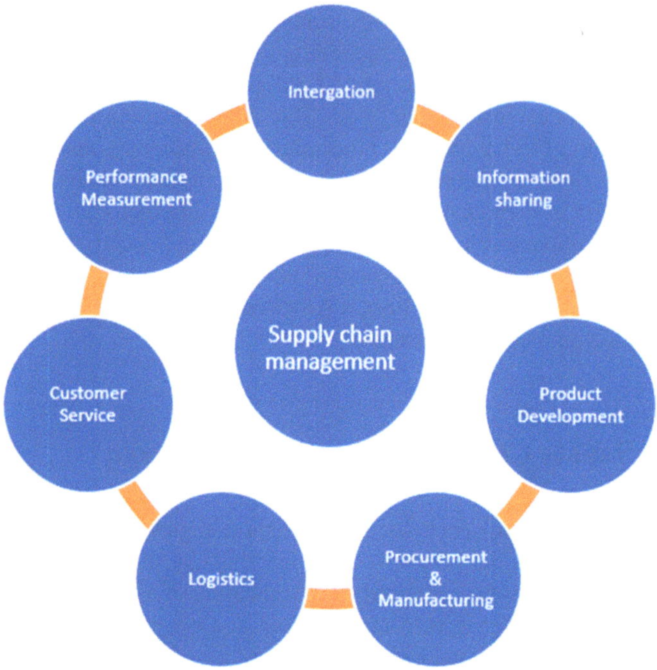

Fig. 1 Overview of supply chain management

- Improved operational visibility;
- Enhanced client services.

The most intruding issue in conventional supply chain management is its transformation to digital context wherein implementing new technology and business models can enhance the data-intensive operations that are not practiced to a greater extent in conventional supply chains [7]. The goal is to increase transparency throughout the whole supply chain to speed up order fulfillment for customers and cut costs for both manufacturers and suppliers.

In the last 10 years, there has been a stupendous growth in supply-chain capabilities and technology, with digital transformation playing a key role [13]. Supply networks have been more automated in recent years. Companies have been able to streamline their supply chains while cutting costs thanks to the deployment of intelligent robots and other technologies.

Companies that are dedicated to supplying chain execution excellence, which aims for maximal utilization of available supply chain resources, manage costs at each stage, and deliver products to customers on time and by specifications, should already be moving in the direction of a digital strategy.

Johnson & Johnson's *planning team has been assigned with developing an advanced full end-to-end infrastructure of the supply chain and enhancing its customer experience by utilizing digital technology to create new experiences. This* *transition is made possible by integrating digital technology into its supply chain operations in several ways, including personalization of service to customers. A digital ecosystem developed analyzes patient's data through predictive analysis and provides personalized treatment advice* [34].

(*This is given as additional information in the context of the digital supply chain*).

2.2 Seven Steps to a Digital Supply Chain Management

1. Make sure you have the fundamentals in order:

If the present setup is having issues with supply chain performance, identify any necessary process adjustments before introducing new technologies.

2. Converge different functions:

Avoid categorizing tasks into separate categories like production and buying. Pay attention to how different stakeholders accomplish the intended business objective. Consider conducting a poll to gather information on needs and gauge awareness.

3. Set goals for your business that go beyond operational effectiveness:

An illustration would be better demand planning. A digital supply chain can respond to queries like Supplier A has missed a purchase order and how could affect our product catalogue by providing a much clear view of the relationship between raw materials and finished goods.

4. Get everybody involved:

I hope your tiger team is ready. You must now broaden the message: Make sure that everyone in the organization, from senior leadership to the shop floor, is aware of the supply chain strategy message. Bringing cross-functional teams can give way to agile development methodology on the technical side.

5. Make a list of your own data:

Consider the supply chain data that the business is already gathering and how it might be improved. For instance, can an algorithm be used to precisely estimate needs rather than manually determining safety stock?

6. Look outside yourself:

Start considering what data you can share with your suppliers and what they can share with you through IoT and API's.

7. Determine the projects that need to be prioritized:

Which ones have the best chance of success? With predictive analytics getting into usage, it becomes easier for greater end-to-end transparency and real-time disruption response The dependability on order processing, smarter logistics planning, warehouse automation, or better demand planning can also be made effective.

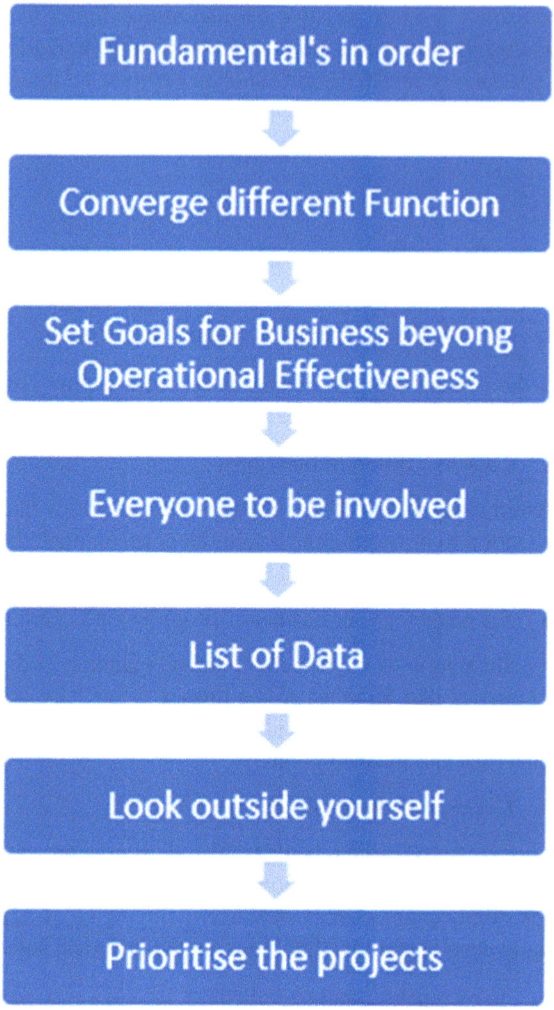

Fig. 2 Seven steps to a digital supply chain

The above-mentioned seven steps as shown in Fig. 2 may help to facilitate easier transformation to a digital supply chain.

3 Digital Supply Chain Management (DSCM)—The Trend

There is currently no "Toyota Way" of the digital supply chain that may serve as a model for other firms to follow, according to an interesting finding made by McKinsey in its analysis. Businesses may be prone to become bogged down by focusing only on one procedure or measure [31]. The project can also become overly IT-focused, lack a solid business champion, and never advance beyond the pilot stage.

However, according to 66% of 234 supply chain professionals who responded to a study by benchmarking firm American Productivity & Quality Center (APQC), integrating new technology and competencies was a high priority prior to the pandemic. Technology is the top area where CFOs are open to extra expenditure.

Building more resilient supply chains is essential, thus these investments should be accelerated given the persistently unpredictable climate. According to Boston Consulting Group, digital supply chains may boost EBITDA, increase revenue by up to 6%, save expenses by up to 20%, and improve customer service by up to 30%. Digital supply chains are at the center of whole new business models as they develop and turn become profit centers rather than cost centers.

Walmart's *recent digital supply chain transformation has made it possible for them to give customers more precise information about when their orders will reach their doorsteps—a capability that was previously only accessible for items purchased directly from customers. A customer can be kept informed of the progress of his or her order up until the time it is delivered to him or her, for example, by using the automated tracking system offered by the provider and thereby making them more involved in the process by incorporating digital technologies into the supply chain.*

(This is given as additional information in the context of Digital Supply chain).

4 Enhancing Digital Supply Chain Through Industry 4.0 (I 4.0)

Disruptive impact due to the pandemic on global supply chain has heightened the need for organizations to focus on risk mitigation and increasing resiliency [13]. Industry 4.0 involves the usage of digital technology for its effective operative functions which has transformed from the advent of smart manufacturing to digitization of entire value delivery channels. Across the globe, various companies are considering the options of integrating their engineering, manufacturing, and planning systems—horizontal integration. On the other hand for vertical integration, they are binding together the other components—hardware, automation equipment, and data lakes. For this to happen, it requires standards and common reference architecture for quick and easier ways of connecting various components. Various technologies in Industry 4.0 assists in this vertical and horizontal integration of supply chain activities which are shown in Fig. 3.

The digitization of the supply chain can boost its resilience by reducing inefficiencies and lower costs while improving flexibility [17]. Table 1 lists the different I4.0 technologies/tools that can aid in effortless transition to digital supply chain. With many technological tools to its support, I4.0 can be that cushion for industries looking for a smooth transition from conventional supply chain to digital mode.

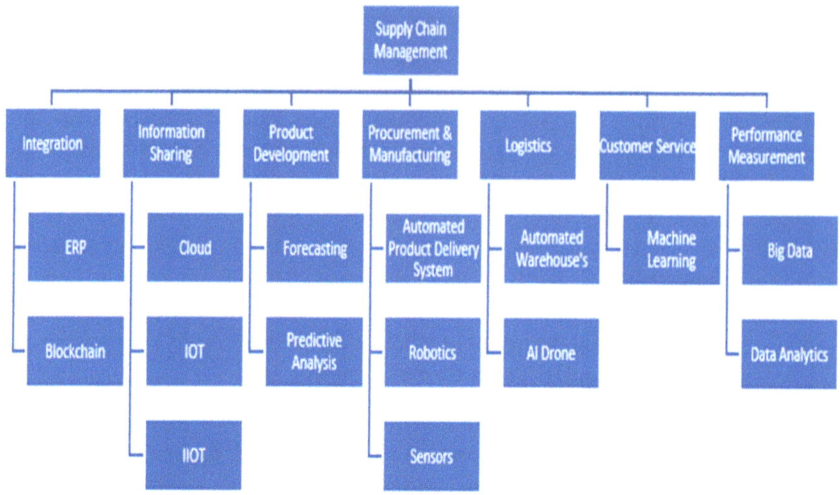

Fig. 3 Industry 4.0 technologies for enhancing digital supply chain

4.1 Big Data

Big data analytics refers to the use of sophisticated analytical methods to enormous data volumes. Big data analytics integrates data and quantitative approaches which improves decisions across the whole supply chain [5]. Beyond the standard internal data kept in ERP systems, it expands the dataset for analysis. Additionally, it uses powerful statistical techniques to examine both recent and historical data sources. Fresh insights are generated as a result, helping supply chain decision-makers with everything from operational decisions like choosing the ideal supply chain operating models to more strategic ones like front-line operations. Businesses that use big data analytics can better understand their consumers' needs, deliver services that satisfy those needs, boost sales and revenue, and diversify into new markets [2, 9].

4.1.1 Top 5 Applications of Big Data Analytics

1. *Relationship Management with Suppliers*

Usage of big data and machine learning can help supply chain partners to effectively organize supply chain activities. Although many companies at present do not employ big data analytics extensively, it is to be firmly believed that future supplier relationship management methods will largely rely on both big data analytics and machine learning [8]. A successful creation of effective relationship management strategies depends mainly on the supplier data which must be reliable and largely qualitative.

Table 1 I4.0 technologies/tools that can aid in effortless transition to digital supply chain

S.no	Industry 4.0 technology/tools	Explanation/Usage	Area of application in SCM
1	Big data analytics	Processing and extracting information from a lot of different datasets. In conventional databases, used for decision-making by combining data quantitative methodologies	Logistics, Inventory
2	Data analysts	Data analysts use this information for creating methods and carrying analytical procedures through which informed decisions can be made with better accuracy. It also assists businesses by saving costs and boost profitability	Demand forecasting, Inventory planning, Quality management
3	IoT	The IoT tools refers to Internet of Things tools. It is an interrelated network of computing devices, mechanical elements and machinery with embedded electronics, home appliances, structures, and more. This facilitates the gathering and exchange of many types of data	Forecasting of product demand, Logistics
4	IIoT	A network of interconnected industrial devices is called the "Industrial Internet of Things." The benefit of IIOT networks' connected devices is that they transmit data without requiring human or computer involvement	Manufacturing, Maintenance, Tracking
5	Sensors	Detect and record various physical attributes of a product and transform them into digital insights which can be used to create new values in supply chain	Manufacturing, Shipping, Retailing
6	Robotics	Robotic systems are defined as those that interact with their environment, including people, using a variety of sensors, actuators, and human interfaces to provide intelligent services and information	Administration, delivery, Warehousing, and inventory control
7	Blockchain	A decentralized, constant database for easier tracking of goods and commodities. It helps to develop digital ledger which can be shared for complete data visibility across the chain in a corporate network	Tracking and Transaction recording
8	Cloud	A cloud based utility is a software that uses cloud application technology to store data centrally and give all users access to the most recent information	Warehouse, purchase, Marketing, etc.

(continued)

Table 1 (continued)

S.no	Industry 4.0 technology/tools	Explanation/Usage	Area of application in SCM
9	Enterprise resource planning	The integrated management of key company processes is known as enterprise resource planning (ERP)	Purchase, Marketing, Production planning control, Purchase schedule entry, Quality, Logistics, Warehouse, etc.
10	Predictive analytics	A part of advanced analytics procedure that uses historical data along with statistical modeling, data mining, and machine learning. to forecast future results. Through predictive analysis, it becomes easier to determine ideal requirement in inventories and minimize stock pile-up at the retailers	Warehouse, production, Manufacturing
11	Automated warehousing	It helps in reducing human intervention at the warehouses where most time is spent on movement into, within, and out of warehouses. Improves supply chain visibility and can assist in automating inventory levels at warehouse	Warehouse
12	Forecasting	The act of forecasting involves making predictions about an industry's demand, supply, and pricing	Purchase, manufacturing, Warehouse, Production
13	Automated product delivery system	The automated product delivery system was created in response to the requirement for an easy method of moving products across a manufacturing environment	Warehousing, Inventory, Management, Transportation
14	AI-Drones and robots	Now, autonomous robots and drones can examine equipment, carry out preventive maintenance, and provide analogue data to the operations crew	Inventory process, Warehousing
15	Machine learning	An application of AI which can predict consumer/market behavior from their past performance and making the supply chain more resilient to disruptions	Predictive Analysis, Quality Inspection, production Planning, Warehouse, etc.

2. *Design and Development of Products*

Big data Analytics can benefit manufacturers by creating plans and sharing data and can benefit them during critical times for decision-making [9]. In addition, big data analytics can help supply chain partners in scheduling order pickup and delivery as well as allocating orders to appropriate agents.

3. *Demand Forecasting*

Orders that are either unfinished or incomplete might harm a company's reputation. In the age of the customer, keeping customers happy and loyal requires providing the ideal product to the right customers at right time and at the ideal location. Smart companies can use big data to get a full analysis of their clients from their perspective which enables them to more accurately predict customer needs, comprehend their preferences, and offer a distinctive brand experience [2].

4. *Logistics Administration*

Optimizing service experiences for logistics systems, such as delivery speed, resource application, and geographic coverage, is a recurring challenge. Both late and early delivery would be costly for logistics companies. The time difference between the planned and actual deliveries is one of the main risks that logistics companies are concerned about. It is advantageous to apply big data analytics in supply chain management to lessen the risk of inaccurate delivery timings [5].

5. *Maintenance of Machines*

Big data is increasingly being used in maintenance, and this development is improving how the maintenance team functions. The move toward data analytics is improving operational uptime by increasing the effectiveness of the maintenance department [8]. The operation can efficiently assess the health and performance of its equipment by combining machine data with other area data. Installing sensors and using the collected data to simulate the device's operation are the main methods used to do this.

4.2 Data Analysts

By using real-time data to simulate failure or defect situations around every conceivable scenario, data analytics helps to identify the underlying causes of problems. It can pinpoint probable sources of mistakes or faults by repeatedly analyzing data from several historical timeframes [3]. As a result, analytics contributes to the extension of quality coverage to upstream manufacturing operations, a crucial component of end-to-end quality improvement projects [10].

With the aid of analytics, producers can find pertinent information on the utilization of raw materials, throughput of machinery and equipment, and operating environment optimization that can improve factory output volume without sacrificing quality. To increase throughput inside a manufacturing plant, analytical tools can provide real-time insights into anticipated production effort and raw material requirements, labor and machine deployment schedules, and coordinated logistics [19]. In furtherance of boosting production output while maintaining quality assurance, data patterns inside processes and workflows can be analyzed to identify the best candidates for optimization, be it of processes, machines, or staff schedules.

By additionally considering the demands and lead times of your organization, inventory management involves using large amount of historical data from sales and forecasting future demand for your inventory. Any type of organization formerly found it challenging to balance product availability versus projected market demand. With the use of data analytics, inventory management has become less complex, and demand forecasting has become possible. Additionally, data analytics offers perceptions of consumer behavior. both the channel's and the product's performance [19]. All of them enable all types of merchants, even those with large databases, to manage everything from stock availability to demand sales to product returns.

4.2.1 Utilizing Data Analytics for Inventory Control

By effectively managing the inventory, data analytics can assist reveal patterns and trends that can be used to enhance corporate operations.

- Increasing operational efficiency—Managers in operations can get a real-time perspective of the operations and a greater understanding of the metrics with the use of data analytics. This helps in eliminating all bottlenecks, enhancing the effectiveness of your sales.
- Efficiency enhancing.
- It helps prevent product overselling and running out of stock.
- Aids in hastening the order fulfillment procedure.
- Maximization of sales and profits.

4.3 IoT

In supply chain management, "IoT" indicates incorporating numerous data points (devices utilized in the supply chain) to produce an optimal result through improved operational efficiency or more accurate demand forecasting for products. Even for tiny businesses, it is a rising habit. IoT connectivity is rapidly expanding in this crucial business practice because it allows for more accurate tracking of items across the supply chain. An IoT-powered supply chain not only helps big enterprises that trade goods internationally but also benefits smaller businesses with fewer resources [4, 28].

That definition is significantly altered by the Internet of Things to include equipment that gathers more data automatically compared to getting data manually.

An IoT supply chain is a system that uses device connection to more effectively distribute a product to a consumer while still achieving the same goals. On the basis of how they connect IoT devices, entrepreneurs gain access to a more dynamic production environment on a smaller scale [28].

4.4 IIoT

IIoT connects a wide range of supply chain linkages to give all parties involved coherence, transparency, speed, and scalability in the development and delivery of their products [11]. Due to globalization and the Internet's ability to bring people closer together, there are now even more levels in the supply chain across all industries. Despite this, businesses continue to confront challenging issues as they attempt to meet these expectations.

1. The assets can be traced using an RFID sensor all through the chain rather than manually logging them in a spreadsheet. Instead of individually scanning the products with a barcode, RFID scanners in each of the factories and warehouses ensure accurate capture of the goods in one batch [17].
2. Warehouse automation enables robots or collaborative robots to collaborate with workers on the warehouse floor. People can take on more complicated, low-level jobs that help the organization since they are programmed to perform a variety of functions and connected to the Internet [27].
3. The data of the materials on each stage of the chain is effectively linked, enabling effective communication among all parties. As well as using real-time information on the location and environment the goods are in, if there are delays in the chain, each level can be contacted to determine what is causing them and where they are occurring. It guarantees safe delivery and track-and-trace models to provide end-to-end live support for clients when they receive their orders and items by working with significant delivery businesses that also employ RFID and IIoT procedures.

4.5 Sensors

Sensors have been widely used for detecting inputs from a physical world like light, heat, motion, wetness, pressure, or any other environmental changes. In the supply chain, sensors help with location, vehicle temperature, vehicle pressure, and more. Supply chain executives may utilize these inputs to track a shipment or vehicle's whereabouts and determine its exact state and destination. The ability of sensors to provide real-time behavior status is what makes them valuable in supply chains (e.g., location and speed of vehicles).

As previously discussed, the usefulness of connecting sensors for the supply chain creates the capacity to provide predictive insights into the process as well as real-time activity [21]. Supply chain executives now have the ability to comprehend the current and future health of their interdependent supply networks thanks to these predictive data produced by IoT sensors.

4.5.1 Benefits of Smart Sensors in the Supply Chain

Smart sensors have given space for supply chain partners to rapidly increase automated data collecting and processing, improve asset efficiency, and provide management visibility throughout the supply chain to assist businesses in lowering operational costs, maximizing asset performance, and generating additional income. Smart supply chain sensors can benefit enterprises by:

- Automate operations to improve efficiency.
- Track inventories in real-time while improving demand planning.
- Encourage the creation of new products and improve product life cycle management.
- Improve customer service through developing a stronger relationship with the client

It is now necessary for businesses to evaluate their supply networks in preparation for testing smart sensors. Smart sensor connections can give managers insights into every stage of the supply chain, resulting in increased efficiency, lower costs, and new income prospects.

4.6 Robotics

Businesses in any part of the world always aim lower their long-term costs and provide labor and utilization stability in their operations. In such scenarios, businesses look for assistance from robots for increasing worker productivity, decreasing error rates, and reducing the frequency of inventory checks. Also, they have been commonly utilized for optimizing picking, sorting, and storing times, and increasing access to challenging or hazardous locations [7].

Robotics have the ability to enhance operations and present fresh chances to boost output, lower risk and lower costs. With the expectations of customers going high every year, conventional methods find difficulty in data collection, especially as consumer expectations and the volume of packages, shipments, and orders reach levels that are varying in every sector. Robots give supportive solutions to these issues by their flexibility to work in diversified places.

4.7 Blockchain

The supply chain can be made more transparent thanks to blockchain, which can also lower costs and risks. Few of the benefits that blockchain supply chain technologies can specifically provide includes:

Primary possible benefits

- Increasing material supply chain traceability will help to ensure that corporate standards are met.
- Lower losses from gray market/counterfeit trading.
- Enhance transparency and compliance with regard to outsourced contract manufacturing.
- Cut back on administration and paperwork expenses.

Additional potential advantages

- Bolster corporate brand by making the materials used in products transparent.
- Boost the credibility and provide wide acceptance of the data shared.
- Activate stakeholders.

A decentralized immutable record of all transactions and the digitization of physical assets by organizations and usage of decentralized record of all transactions enhance the possibility to track assets from manufacture to delivery or end-user use. This has enabled more transparency and precise tracking of end-end customer services in supply chain [24]. With this improved transparency, it gives a win-win situation for both businesses and consumers as they now have a better visibility into the supply chain [6].

Few companies which have outsourced their production will have more control over production flow by using blockchain for reducing errors in communication or data transfers as it gives all participants of an active supply chain the access to same information [20]. This could speed up the data verification process and utilize the available time for delivering goods and services by raising quality, lowering costs, or doing both.

Finally, by enabling a successful audit of supply chain data, blockchain can improve administrative processes and lower expenses.

4.8 Cloud

Cloud enables tasks to be automated in a business, gain fresh insights for decision-making, and innovate processes at every stage of their supply chain. from planning and sourcing through production, transportation, and distribution. Supply networks are made more robust and intelligent as a result [12].

A connected, intelligent, scalable, and quick supply chain made possible by the cloud gives enterprises unprecedented visibility, robustness, agility, and flexibility.

(1) *A Linked Supply Chain*

Connectivity of the supply chain network improves through cloud solutions by utilizing the structured and unstructured data generated throughout the supply chain. This makes it simpler for businesses to communicate with their clients, vendors, and suppliers. Additionally, the cloud makes it possible for datacenter efficiency, greater server usage, and dynamic provisioning.

The integrated supply chain's advantages.

- instantaneous visibility,
- Effortless cooperation,
- modern operating systems.

(2) *A Smart Supply Chain*

For every business, getting right information at the apt moment is essential for ideal decision-making, which cloud computing provides through its cognitive intelligence and predictive analytics. Potential dangers are identified through analytics, and the corresponding decision-making is done proactively through cognitive intelligence. Cloud technology provides useful information that enable sophisticated digital supply networks.

Advantages of an intelligent supply chain include:

- shortened period of disruption mitigating,
- better automation,
- increased inventiveness.

Supply chain that is scalable: Scaling up and down the supply chain network depending on consumer demands and market conditions is made much easier as the cloud solutions operate on a flexible, usage-based paradigm making the network and storage required to be more flexible to the given change.

(3) *The Scalable Supply Chain's Advantages*

  - maximal effectiveness,
  - business-level adaptability,
  - increased client satisfaction, and
  - reduced costs.

4.9 ERP

The importance of the ERP system for SCM can't be hyped. This software has significantly turned how to operate in the industry. This effective system was crucial for corporate development. The SCM technology was acquired from professionals like QAD, Merge the SCM operation into a particular cell, enhancing visibility and streamlining cooperation between vendors and suppliers [23]. In the production unit, ERP system finds out the lagging area and is used to improve it.

The ERP system calculates the market demand and plans, It produces when orders are getting. which is calculated by the program when the order is placed. In order to boost performance, it is possible to automate or streamline the administration of inventory resources, and other supply chain procedures [24].

Although there are on-premises ERP system options, cloud ERP software has grown in popularity recently. Mobile access and dedicated data security are two of cloud ERP's main advantages. If you use an ERP solution that is hosted in the cloud, you can access the system from any location at any time using a mobile device if you have an internet connection. Due to the fast-moving nature of supply chains, this enables rapid action without necessitating the presence of the entire team.

4.10 Predictive Analytics

Businesses may now choose the ideal sales volumes to boost supply while reducing inventories thanks to data analytics. Supply chain managers may determine the precise inventory needs by region, demand, and region using analytical techniques. Safety stock can be reduced as a result, and inventory can be placed where it is needed [25].

Logistics and shipping:

The cost of transportation is often included in the price of the finished product. Forecasting can be used to establish the optimal supply frequency and amount to fulfill demand and lower costs [29]. Proactive route planning can help choose the optimal routes by taking delivery locations, location, climate, and busy highways into account. Furthermore, intelligent items may continuously monitor things like air density, driving habits, and vetting.

Strategies for supply chain preventive maintenance:

The requirement for service can be foreseen with prospective site monitoring, which also provides ample notification of system problems. By keeping more outdated components on hand, businesses can save on inventory and prevent unanticipated equipment breakdowns by using this information to place item orders as needed. Using the Airbus Points Management program, which regularly checks on Eastern Air Lines aircraft and keeps track of the state of the gear, is a great example. This enables the airline to schedule servicing on the aircraft in advance and decrease unplanned downtime.

4.11 Automated Warehousing

The automated processing of inventory coming into, going through, and leaving warehouses for delivery to customers is known as Industry 4.0. Automation enables a company to do away with labor-intensive processes including physical entering data, analytics, and ongoing backbreaking labor.

Distribution network robotics is the use of digitalization to activities in the supply chain to improve efficiency, integrate apps, and expedite processes. Emerging methods including AI technology, algorithms, automation of robotic procedures, and virtual intelligent systems are often used.

A linked distribution network supported by Industry 4.0 might relieve workers of these tiresome tasks.

4.12 Forecasting

The act of estimating consumption, supply, or viability for a resource or a selection of goods in a certain industry is known as supply chain projection.

For instance, a prediction model's computations can predict a sales volume by looking at data from vendors and consumers [1]. The computer may also take into account external factors like the environment or other disrupting circumstances to increase the costing estimate's precision.

Merchandise forecasting is the strategy of estimating how often stuff you might acquire over a certain amount of time, and it is used to determine how much resources organization will be required to satisfy rising client wants. To be as precise as possible, our projections consider for historical sales information, anticipated bonuses, and outside factors.

4.13 Automated Product Delivery System

Automated processes are the overseeing of the movement of commodities into, within, and out of warehouses for delivery to customers. As part of an optimization effort, a corporation can cut the number of employees performing manual data input, repetitive physical labor, and analytical work [22].

For illustration, a warehouse worker may load bulky objects onto a moving robot system. The software keeps all data up to date while inventory is drone-moved from one section of storage to the delivery zone. These robots boost the effectiveness, speed, reliability, and accuracy of this task.

The use of technology to replace a labor-intensive manual system is commonly referred to as "industry 4.0," but the technology need not be physical or mechanical.

4.14 AI-Drones and Robots

Without the assistance of a human controller, a robot can reach out and grasp an object thanks to AI. Motion control and navigation increased by AI. Robots become more autonomous thanks to improved machine learning capabilities, which eliminates the need for humans to organize and control process flows and navigation paths. The future of warehouse management is going to be done only by the robots in future [26].

Drones can fly, carry a certain amount of weight, move objects, and access inaccessible places (this includes geographic areas, like isolated villages, and deep shelves or places that are higher in a warehouse).

Drones in inventory

Drones are used in warehouses to complete activities including stock searching, cycle counting, and inventory audit. They fly independently, recognize and count the inventory held in the warehouse, and compare it with the data recorded "virtually" to enable safe and cost-effective inventory operations.

Drones in logistics

Unmanned delivery is already a reality which has been used recently through Amazon which provided much easier and quicker delivery of products to the customers. Commercial drone delivery services are timelier and more cost-effective than traditional delivery services.

4.15 Machine Learning

Usage of machine learning algorithms has made the supply chain managers to optimize the route for their fleet of vehicles by analyzing and learning from real-time data and historical delivery records, which has resulted in reduced driving time, cost savings, and higher productivity.

ML in inspection

The machine learning testing technique offers accuracy and efficiency advantages in the quality assurance process. It helps with the identification of redundant and unsuccessful test cases as well as the forecasting and averting of code issues. Inspections carried over are uploaded in real-time to the customized digital platform. ML algorithms provide advanced analyzation and to the given data, which provides greater insights into the process which is not possible using legacy manual methods.

ML in production

Continuous improvement is a feature of machine learning. Therefore, in demand planning, the machine learning engine assesses the model's forecast accuracy and considers if the forecast would be better if the model were altered in some way. Forecasts are continuously and iteratively improved.

5 Industry 4.0: How Digitization Makes the Supply Chain More Efficient and More Effective

Most business processes need to be more digitalized if the Industry 4.0 goal is to be achieved. It is this transformation of conventional supply chains which has become a crucial factor. The change into an ecosystem of interconnected, intelligent, and highly effective supply chains will make the ecosystem more effective and efficient [30].

The new age supply chain has different phases on its own that start with marketing, continue through product development, manufacturing, and distribution, and end with the customer receiving the goods. Some hurdles which are once considered disrupting the chain have been broken down by digitization. The entire flow of chain has been made into a fully integrated ecosystem that is transparent to all of the participants, starting from the purchase of raw materials, components, and part suppliers to the carriers of those supplies and finished goods.

The current technologies and tools exponentially increase the supply chain network functions. The re-modeling of the supply chain network by developing what-if cases and with real-time data would result in an increase in responsiveness to any condition and to anticipate them. This will give pathway to digital supply chain with more resilience and responsiveness instantly to the demand of the customer which allows to beat the competition and enable system transparency. The research would enable us to see how these would pave the way for digital supply chain.

6 Value Drivers for Digital Supply Chain Management (DSCM)

The current technologies disrupt the value chain; organization has to look for sustainable ways to design their supply chain and how to integrate the latest tools. There are some visible drivers such as big data and optimization tools, there are other few value drivers which will push towards digitization of supply chain. Few of the key value drivers and their respective technologies/tools that drive businesses towards digitization of supply chain are listed in Table 2.

1. Collapsing product lifecycles

The product lifecycle has been reduced by 50% in the last decade and would reduce even more in the recent upcoming years. This is mainly due to the upcoming of new product in the market and replacing within two years as per the norms of the industry. If a company doesn't come up with a new product as soon as possible, it would risk losing the market or beaten by a competitor.

Table 2 Key value drivers and their respective technologies/tools driving towards digitization of supply chain

Value drivers	Technologies/tools
Supply chain strategy	Dynamic network configuration
Collaboration	Cloud supply chain, End-to-end connectivity
Order management	Reliable online monitoring, real-time planning
Performance management	Automated root cause analysis, Online Transparency
Physical flow	Warehouse automation, Autonomous smart vehicles, Human–machine interface, smart logistics, 3D planning
Planning	Predictive analysis in demand planning, closed-loop planning, and scenario planning

2. Outsourced Manufacturing

The latest supply chain strategy is to outsource the product of different variants so as to decrease the level of risk. This enables the company to concentrate more on the core product of the company. Also, this would strategic advantage for supplying and manufacturing.

3. Generational Expectations

In the current era people expect everything to be unique and customized to satisfy their needs that is they research their need, sustainable product, embrace sharing model, and instant access to customized product.

4. Supply Chain Savvy in the C-Suite

The supply chain management change in the current era requires digitization as this would give a competitive edge with their competitors in the global market. The current supply chain model has outgrown the conventional model to encounter the demand by utilizing flexibility, actionable insight, and real-time data with the help of digital platforms.

7 Conclusion and Future Work

International markets getting widened up and being globalized and penetrated into international boundaries, it has become utmost necessary for businesses to survive and to maintain a competitive advantage and attain sustainability in global market. They, by all means, have to identify suitable identify emerging digital technologies that can be used to develop a new business model for their organization. This chapter throws light on principles and practices for digitized SCM, and the technologies/ tools that drive the firm towards digitized SCM. The corresponding areas of supply chain wherein they can be applied to utilize its potential for a digitized platform of SC activities have been elaborated, enabling firms and other researchers in deploying further conceptual and empirical works linked to the subject herein explored. This work expedites identification of important themes and areas for practitioners to look for modes of implementation of Industry 4.0 standards in supply chains. Future researchers needed to develop more real-time models that will assist in easier implementation of Industry 4.0 all through the chain. As real-time visibility all through the supply chain has become a necessity, long-term collaboration between the different stages of the chain through I4.0 technologies can drive them to have a fully digitized platform for all SC activities. For future researchers and SC practitioners, this article adds valuable insights, by providing them a platform to explore the upcoming and unexplored concept of the Industry 5.0 phenomenon in the supply chain context, which could extensively provide greater supply chain solutions in future.

References

1. Buyukuzkan G, Goçer F (2018) Digital supply chain: literature review and a proposed framework for future research. Comput Ind 97:157–177
2. Nasiri M, Ukko J, Saunila M, Rantala T (2020) Managing the digital supply chain: the role of smart technologies. Technovation 96:102121
3. Treiblmaier H, Mirkovski K, Lowry PB, Zacharia ZG (2020) The physical internet as a new supply chain paradigm: a systematic literature review and a comprehensive framework. Int J Logist Manag 31(2):239–287
4. Garay-Rondero CL, Martinez-Flores JL, Smith NR, Morales SOC, Aldrette-Malacara A (2020) Digital supply chain model in Industry 4.0. J Manuf Technol Manag
5. Radivojević G, Milosavljević L (2019) The concept of logistics 4.0. In: 4th Logistics international conference, pp 283–292
6. Bamberger V, Nansé F, Schreiber B, Zintel M (2017) Logistics 4.0–facing digitalization-driven disruption. Prism 38:39
7. Amr M, Ezzat M, Kassem S (2019). Logistics 4.0: definition and historical background. In: 2019 Novel intelligent and leading emerging sciences conference (NILES), vol. 1, pp 46–49. IEEE
8. Agrawal P, Narain R (2018) Digital supply chain management: an overview. In: IOP conference series: materials science and engineering (vol. 455, no. 1, p 012074). IOP Publishing
9. Queiroz MM, Pereira SCF, Telles R, Machado MC (2019) Industry 4.0 and digital supply chain capabilities: A framework for understanding digitalisation challenges and opportunities. Benchmarking Int Journal.
10. Ageron B, Bentahar O, Gunasekaran A (2020) Digital supply chain: challenges and future directions. Supply Chain Forum Int J 21(3):133–138, Taylor & Francis
11. Korpela K, Hallikas J, Dahlberg T (2017) Digital supply chain transformation toward blockchain integration. In: Proceedings of the 50th Hawaii international conference on system sciences
12. Iddris F (2018) Digital supply chain: survey of the literature. Int J Bus Res Manag 9(1):47–61
13. Ivanov D, Tsipoulanidis A, Schönberger J (2019) Digital supply chain, smart operations and industry 4.0. In: Global Supply Chain and Operations Management. Springer, Cham, pp 481–526
14. Ivanov D, Dolgui A (2021) A digital supply chain twin for managing the disruption risks and resilience in the era of Industry 4.0. Prod Plann Control 32(9):775–788
15. Attaran M (2020). Digital technology enablers and their implications for supply chain management. Supply Chain Forum Int J 21(3):158–172, Taylor & Francis

16. Gupta S, Modgil S, Gunasekaran A, Bag S (2020) Dynamic capabilities and institutional theories for Industry 4.0 and digital supply chain. Supply Chain Forum Int J 21(3):139–157, Taylor & Francis
17. Hanaysha JR, Alzoubi HM (2022) The effect of digital supply chain on organizational performance: an empirical study in Malaysia manufacturing industry. Uncertain Supply Chain Manag 10(2):495–510
18. Farahani P, Meier C, Wilke J (2017) Digital supply chain management agenda for the automotive supplier industry. In: Shaping the digital enterprise. Springer, Cham, pp 157–172
19. Meier C (2016) Digital supply chain management. In: Digital enterprise transformation Routledge, pp 231–262
20. Raji IO, Shevtshenko E, Rossi T, Strozzi F (2021) Industry 4.0 technologies as enablers of lean and agile supply chain strategies: an exploratory investigation. Int J Logist Manag
21. Nagar D, Raghav S, Bhardwaj A, Kumar R, Singh PL, Sindhwani R (2021) Machine learning: best way to sustain the supply chain in the era of industry 4.0. Mater Today Proc 47:3676–3682
22. Ghadge A, Kara ME, Moradlou H, Goswami M (2020) The impact of Industry 4.0 implementation on supply chains. J Manuf Technol Manag
23. Rasool F, Greco M, Grimaldi M (2021) Digital supply chain performance metrics: a literature review. Meas Bus Excell 26(1):23–38
24. Zekhnini K, Cherrafi A, Bouhaddou I, Benghabrit Y, Garza-Reyes JA (2020) Supply chain management 4.0: a literature review and research framework. Benchmarking Int J
25. Núñez-Merino M, Maqueira-Marín JM, Moyano-Fuentes J, Martínez-Jurado PJ (2020) Information and digital technologies of Industry 4.0 and Lean supply chain management: a systematic literature review. Int J Prod Res 58(16):5034–5061
26. Barbieri P, Ellram L, Formentini M, Ries JM (2021) Guest editorialEmerging research and future pathways in digital supply chain governance. Int J Oper Prod Manag
27. Da Silva VL, Kovaleski JL, Pagani RN (2019) Technology transfer in the supply chain oriented to industry 4.0: a literature review. Technol Anal Strateg Manag 31(5):546–562
28. Marmolejo-Saucedo JA (2020) Trends in digitization of the supply chain: a brief literature review. OPENAIRE
29. Chauhan C, Singh A (2019) A review of Industry 4.0 in supply chain management studies. J Manuf Technol Manag 31(5):863–886
30. Frederico GF (2021) From supply chain 4.0 to supply chain 5.0: findings from a systematic literature review and research directions. Logistics 5(3):49
31. Anne-TitiaBove and Steven Swartz (2016) www.mckinsey.com/media/McKinsey/Business%20Functions/Sustainability/Our%20Insights/Mapping%20the%20benefits%20of%20a%20circular%20economy/Mapping-the-benefits-of-a-circular-economy.pdf. Accessed 28 Feb 2023
32. Wodecki B (2021) https://www.datacenterknowledge.com/author/Ben-Wodecki. Accessed 20 Feb 2023
33. Sahi S (2022) https://noise.getoto.net/author/snigdha-sahi/. Accessed 24 Feb 2023
34. Ting (2018) Johnson & Johnson: embracing digital transformation. Accessed 24 Feb 2023

Supply Chain Decision-Making Using Artificial Intelligence and Data Analytics

Lukman A. Akanbi, Kayode I. Adenuga, and Hakeem Owolabi

Abstract This chapter examines the use of artificial intelligence, data analytics and other digital technologies in the management of the supply chain decision-making. The study highlights the challenges faced by supply chain managers and how the application of AI and data analytics can help in making better and more informed decisions with respect to sustainability. Data analytics, AI techniques, such as machine learning, natural language processing and other digital technologies that include Internet of Things, Robotics and Cloud computing and their applications to different areas of supply chain management, such as demand forecasting, inventory management and logistics optimisation are discussed. Some of the challenges (initial cost of physical and cloud resources, change management, ethical and legal-related issues) that the supply chain managers need to put into consideration when adopting these technologies are also presented. The chapter concludes that continuous data collection and storage across all the stakeholders in the supply chain must be ensured to enable transparent and efficient use of AI algorithms to support quick and timely supply chain decision-making.

Keywords Artificial intelligence · Data analytics · Supply chain management · Decision-making

L. A. Akanbi (✉)
Big Data, Enterprise and Artificial Intelligence Laboratory, University of the West of England, Bristol, UK
e-mail: Lukman.Akanbi@uwe.ac.uk

K. I. Adenuga
Department of Digital Technologies (Pearson) Global Banking School, Birmingham, UK

H. Owolabi
Greater Manchester Business School (GMBS), University of Bolton, BL3 5AB, Bristol, UK
e-mail: h.owolabi@bolton.ac.uk

K E K Vimal et al. (eds.), *Industry 4.0 Technologies: Sustainable Manufacturing Supply Chains*, Environmental Footprints and Eco-design of Products and Processes,
https://doi.org/10.1007/978-981-99-4819-2_2

1 Introduction

Nowadays, supply chain decision-making is becoming increasingly complex [6]. As businesses strive to remain competitive and efficient, they must be able to make decisions quickly and accurately in order to stay ahead of the competition. Artificial intelligence (AI) and data analytics are two powerful technologies that can help organisations to make better-informed decisions about their supply chains [14, 15]. AI-based systems can process large amounts of data quickly while providing insights into trends and patterns that may not have been previously noticed. By leveraging AI technology, companies can gain insight into customer demand for goods or services as well as analyse past performance metrics such as delivery times or inventory levels to enable optimisation of future operations accordingly. In addition, AI-driven predictive analytics allows companies to anticipate potential problems before they arise by predicting possible outcomes based on current market conditions and historical data points [44]. Data analytics also plays a critical role in effective supply chain decision-making by helping organisations to identify areas where improvement is needed within their processes or systems so that resources can be allocated appropriately for maximum efficiency gains throughout the entire operation cycle from procurement through distribution channels all the way down to the final sale at point-of-sale locations worldwide [4]. Through advanced analytics techniques such as machine learning algorithms which automate tasks such as forecasting demand patterns over time, businesses are able to take advantage of real-time insights regarding pricing strategies, product availability and other important factors leading them towards improved profitability. By utilising both AI and Data Analytics together, enterprises are well-equipped with necessary information required for successful supply chain decision [25].

As the world population continues to grow and projected to reach 8.6 billion by 2030 [45], the demand for goods and services would also increase as a consequence of the population growth. Therefore, the need for manufacturing organisations to embrace sustainability in their operation cannot be overemphasised because, as the need for goods and services increase, the resources required for their production remain limited. According to Bag et al. [4], the only way to tackling the social, economic and environmental problems associated with the increased production is through the towing of the path of sustainable development, by ensuring that responsible production and consumption patterns are maintained. While big organisations at the centre of goods and services production and delivery are showing commitment to sustainability in their operation, the commitment of tier 1 suppliers and others in the supply chain vertical network are difficult to evaluate. Data analytics, AI and other digital technologies can provide the much-needed technology ecosystem to enable all the actors in a supply chain to transparently engage sustainability in their operation.

2 Sustainable Manufacturing

Sustainable manufacturing is an important concept in today's global economy. It refers to the practice of goods and services production in an environmentally friendly way and at the same time ensuring socio-economic benefits [28]. Sustainable type of production involves waste minimisation, energy consumption reduction, use of renewable materials and ensuring safe working conditions for employees [3, 28]. The United States Environmental Protection Agency (EPA) report gave a more succinct description of sustainable manufacturing as the production of goods and services using energy-efficient and economically sound processes that are with less negative impacts on the environment [18]. The benefits of sustainable manufacturing extend beyond just the environment [21, 28], organisations can reduce their carbon footprint while simultaneously increasing efficiency through improved processes such as efficient resource utilisation and better product design practices [38]. Additionally, sustainable manufacturing can help companies save money by cutting down on costs associated with pollution control and disposal fees for waste and hazardous. Business may also be eligible to receive tax credits from governments if they make efforts towards sustainability initiatives like recycling programmes or green building strategies which further incentivises companies to adopt sustainable practices within their operation materials [2, 40].

The advantages associated with sustainable manufacturing have made it to become increasingly popular among businesses. These benefits make sustainable manufacturing an appealing option for any company looking to stay competitive in nowadays marketplace while still preserving the environment at large. To achieve sustainable manufacturing, a sustainable supply chain is a prerequisite where all processes undertaken by the supply chains' stakeholders are sustainable. Both suppliers and customers in the supply chain networks must transparently embrace sustainable practices.

3 Sustainable Supply Chain Management

Sustainable development is described as the process through which the present meets their need through the manufacturing of goods and services with a conscious effort to allow the future generations to meet their needs also [30, 37]. To expand the discussion, sustainable manufacturing refers to the development of environmentally and socially friendly systems to process raw materials into economically viable goods. Increased population with high demand for industrialisation has continued to create increasing pressure on the planet, and the need for sustainable development continues to be the greatest challenge in human history [13]. The manufacturing sector has a major role in ensuring a sustainable society. The motivations for doing so are closely linked to cost reduction as there was a dramatic increase in material and energy inputs, including disposal costs that deplete non-renewable resources over the last decade [7]. Thus, increased environmental degradation has resulted in stricter legislation, fines for non-compliance and public campaign initiatives that have driven companies toward creating socially and environmentally friendly products and services [12].

Sustainable manufacturing supply chain management is an all-encompassing concept that has recently sparked a growing body of knowledge and interest. While the term supply chain deals with the coordinated activities of all companies' resources, aggregated towards developing, manufacturing and delivering a product, supply chain management refers to material sources coordination, production, inventory management and transportation to maximise efficiency and customer experience [7, 38]. Thus, supply chain management ensures that companies comply with the environmental standards and societal values which includes tackling global challenges, i.e. climate change variables and related social issues and combining all the related components. The goal of sustainable supply chain is to reduce environmental harm that could result from various activities of the supply chain actors. These concerns are in addition to the traditional corporate supply chain concerns around revenue and profit [27]. Recently, companies and businesses have faced an increasing need to focus on the implications of the business activities of their supply chain on both the social and environmental components of society. [22]. They are being held accountable to sustainable utilisation of resources [10]. Supply chain management ensures cost control, customer experiences, quality, challenges and opportunities [36]. It focuses more on sustainable business models than product delivery [5]. Most challenges with the supply chain do not occur at the final stage of the product but during the process of sourcing, production, delivery and logistics. Thus, it suffices to say that all activities that took place in the supply chain constitute threats in some ways to society in the form of environmental, social and economic threats. [39]. A growing number of works conclude that over 90% of carbon-related emissions are caused by activities related to the supply [41], sparkling researchers' renewed interest in the area of sustainable supply chains to curtail the environmental and social threats posed by the supply chain activities. Supply chain activities' by-products include electronic waste, pollution (particulate matters) and greenhouse gas emissions [1]. Thus, it becomes imperative to ensure that an effective mechanism is put in place to manage the challenges within the operational and marketing framework of the supply chain [39]. Although companies have adopted various strategies to tackle challenges impeding efficient supply chain networks [24]. There have been calls to ensure that a more robust and responsive supply chain is created through artificial intelligence and big data to ensure effective monitoring, forecasting and optimisation [36]. Although businesses have started investing in, and adopting AI technologies for improving the operation of their supply chains' end-to-end activities [11, 23], more needs to be done to embrace the use of the latest development in data analytics and AI for maximum benefits and good return on investment. Natural language processing, machine learning and robotics are some of the AI technologies that have been demonstrated as potential enablers of supply chain transformation [36].

4 AI and Data Analytics in Supply Chain Management

Artificial intelligence is a branch of computer science that deals with the creation of intelligent systems that perform tasks ordinarily meant for only human to undertake [35]. It is a field that impersonates human capabilities to understand their environment, process information and makes decisions to achieve a set goal. AI has aided decision-making in many sectors, including banking, farming and autonomous systems. For instance, AI and digital technologies have optimised land use by farmers to yield large harvests and ultimately enhance biodiversity [9, 29, 31]. Studies have shown that the AI technologies' share of the global economic stands at circa £11 trillion by 2030, thereby leading to a 1.2% increase in the global GDP per year [8]. AI has enjoyed massive patronage from businesses, including IBM Watson, (used in diverse areas from medical diagnosis to insurance), Fleet learning system developed by Tesla (which makes it easy for vehicles to continuously share their changing condition data with the rest of the fleet).

In recent times, the advancement in AI and other innovative digital technologies has made it possible to surmount many challenges in the supply chain [33]. There is a growing trend to achieve sustainability in business by applying many digital technologies that include Internet of Things (IoT) for real-time data collection from the field, big data analytics for data extraction, transformation, loading, storage and insights generation and AI for predictive model development [39]. Moreover, the application of AI and other digital technologies has progressed in wider areas of endeavour [34], it has allowed tasks to be automated and decisions made autonomously by machines [36]. This is so because AI can equip machines with human-like reasoning capabilities [43]. Some of the challenges relating to speed and accuracy can be solved with digital technologies, especially when larger inputs are involved. Evidence shows that many companies across the globe have integrated digital applications into their supply chain management [26]. Many corporations are now using machine learning techniques of AI to manage product development through intelligent distribution of information to facilitate sustainable supply chain management [32]. In the same vein, supply chain managers have started focusing on the use of AI to aid decision-making by using big data analytics to predict supply chain uncertainties more accurately [20]. Although the potential use of AI has increased, the successful implementation poses several challenges such as high initial cost of hardware, software and cloud infrastructure. Other challenges include training requirements and change management of personnel involved in the supply chain processes being automated, ethical and legal issues for all the stakeholders. Therefore, for AI to be successfully applied in the management of sustainable supply chain, it is essential to have a more comprehensive view of the process and ensure that all the macroenvironmental factors are considered to assure high confidence among all the stakeholders. The following subsections discuss the role of AI and its associated technologies in managing the supply chain in the manufacturing sectors and subsequently discuss benefits and implications.

4.1 *Potentials of AI and Data Analytics in Supply Chain Management*

The use of AI and associated technologies have aided companies in serving their customers better through the creation of a more efficient, automated and collaborative supply chain system. The benefits derivable from the adoption of AI and digital technologies are twofold; (1) internal benefits derived from disruptive technologies to aid supply chain management and (2) customer satisfaction benefits. Besides, most emerging digital solutions help to mitigate the socio-environmental risks associated with the supply chain operations and by implication thereby offering increased sustainability. Blockchain technology, for instance, provides a translucent process that offers increased consumers' trust on product's quality and supply chain visibility as well. New technologies' application in the supply chain would produce a huge dataset that can be used to develop predictive and prescriptive AI models and advance analytical methods that can assess the impacts of the new technology on supply chain management.

Cloud computing is another digital technology that offers operational intelligence in supply chain management. With cloud computing infrastructure, companies can gather real-life consumer and supplier data that can reveal customer and supplier requirements, especially in creating a match between demand and supply. Robotics, on the other hand, has played a significant role in supply chain management and sustainability. With the proliferation of autonomous and collaborative robots, companies now have access to new opportunities that can reduce costs in terms of labour and productivity (B2be, 2022), error rate reduction, sorting and storage time optimisation. Over the years, the impressions people have about the robot's primary function is to replace humans in doing all complex activities; however, the purpose of robots today (i.e. collaborative robots (cobot)) is to collaborate with humans to reduce the time to carry out a specific task. Robotic systems now exhibit autonomy with the ability to be context-aware and act accordingly since they are equipped with advanced sensors which coordinate their movements in relation to their environment. Cognitive robotics is a concept that some companies have adopted, and it will continue to grow significantly in supply chain management. The presence of expert technicians is no longer required for the industrial robots to be programmed and installed as the end users can manipulate these programs independently. Most warehouses are becoming increasingly automated, and autonomous mobile robotics is becoming prevalent. As such, in supply chain management, robots are expected to be natural collaborators but not an alternative to human [16, 19]. Robotic process automation helps supply chain management automate tasks, streamline operations and eliminate human error. It allows supply chains to speed up and meet supply as demand escalates. Robotics also allows greater workforce efficiency, provides safety for workers by taking over complex tasks and improves delivery times.

4.1.1 Supply Chain Decision Support System

In the past years, statistical analysis of the key performance indices of the supply chain's demand planning and forecasting was the only form of data analytics available. Excel spreadsheets were usually used for storing and processing the supply chain data from different stakeholders. By 1990, enterprise resource planning (ERP) and electronic data interchange (EDI) systems were used by organisations for supply chain data and information processing. The ERP and EDI systems support businesses by providing easy access to data that aids in designing, planning and forecasting. In 2000, businesses became more intelligent and gave birth to predictive analytics software solutions. Thus, predictive analytics software assists corporations in gaining deeper insights into how their supply chains function and consequently aided decision-making and network optimisation. Although this has some setbacks since a large amount of data is expected to be generated, there is growing concern about how this large amount of data can be best utilised in the supply chain network. For instance, in 2017, it was estimated that more data (50 times and above) were accessed by the supply chain companies than in the period between 2011 and 2016 together [17], but more than three-quarters of the data remained unanalysed.

According to [46], supply chain data is usually composed of more than 80% of unstructured data and about 20% of structured and easy to analyse data. Thus, organisations' primary challenge lies in analysing unstructured data. However, recent studies identify AI as the game-changing technology in supply chain management. AI is not limited to processing and retaining information but also, it can reason, learn and behave like a human. It can process a large amount of structured and unstructured data and provide a summary of the analyses instantly. AI does not only correlate and interpret data across various sources in the network. It also enables companies to analyse real-time supply chain data to generate required insights and intelligence. The AI technique that is at the centre of the possibilities mentioned above is Machine Learning (ML) which combines the power of algorithms, data science and software to learn without any particular programming. ML models learn to understand trends and patterns and identify irregularities over time before rendering predictive insights. Once patterns are detected through frequently generated data, ML model can suggest the next line of actions that can prevent final breakdown, especially in the supply chain companies, likewise having the capacity to predict upcoming faults and route optimisation in the supply route channel. The challenges of unstructured data are handled by the AI through the use of supervised, unsupervised and reinforcement learning as appropriate for the development of the ML models. AI and data analytics technology can help supply chain managers make quick and informed decisions that ensure stock availability and map transportation routes to reduce idle time and fuel consumption. AI can also highlight areas for improvement, especially in areas where humans might have missed, thereby drastically reducing any future adverse occurrence [42].

5 Conclusion

As the clamour for sustainable development penetrates the manufacturing industry, the need to ensure that manufacturing supply chains are sustainable becomes more important. Supply chain managers should embrace AI and its associated technologies for managing their supply chains. Concerted effort must be put into data collection and storage by all the stakeholders in the supply chain to enable transparent and efficient use of AI algorithms to support quick and timely decision-making.

References

1. Akan MÖA, Dhavale DG, Sarkis J (2017) Greenhouse gas emissions in the construction industry: an analysis and evaluation of a concrete supply chain. J Clean Prod 167:1195–1207
2. Badurdeen F, Jawahir IS (2017) Strategies for value creation through sustainable manufacturing. Proc Manufact 8:20–27
3. Bag S, Pretorius JHC (2022) Relationships between industry 4.0, sustainable manufacturing and circular economy: proposal of a research framework. Int J Organ Anal 30(4):864–898 https://doi.org/10.1108/IJOA-04-2020-2120
4. Bag S, Wood LC, Xu L, Dhamija P, Kayikci Y (2020) Big data analytics as an operational excellence approach to enhance sustainable supply chain performance. Resour Conserv Recycl 153:104559
5. Belhadi A, Kamble S, Fosso Wamba S, Queiroz MM (2021) Building supply-chain resilience: an artificial intelligence-based technique and decision-making framework
6. Bode C, Wagner SM (2015) Structural drivers of upstream supply chain complexity and the frequency of supply chain disruptions. J Oper Manag 36:215–228. https://doi.org/10.1016/j.jom.2014.12.004
7. Bogue R (2014) Sustainable manufacturing: a critical discipline for the twenty-first century. Assem Autom 34(2):117–122. https://doi.org/10.1108/AA-01-2014-012
8. Bughin J, Seong, J, Manyika J, Chui M, Joshi R (2018) Notes from the AI frontier: modeling the impact of AI on the world economy. McKinsey Global Inst 4
9. Cambra Baseca C, Sendra S, Lloret J, Tomas J (2019) A smart decision system for digital farming. Agronomy 9(5):216
10. Costa J (2021) Carrots or sticks: which policies matter the most in sustainable resource management? Resources 10(2):12
11. Chui M, Henke N, Miremadi M (2019) Most of AI's business uses will be in two areas McKinsey Analytics. https://www.mckinsey.com/%7E/media/McKinsey/Business%20Functions/McKinsey%20Analytics/Our%20Insights/Most%20of%20AIs%20business%20uses%20will%20be%20in%20two%20areas/Most-of-AIs-business-uses-will-be-in-two-areas.ashx. Accessed 2 Jan 2023
12. Despeisse M, Mbaye F, Ball PD, Levers A (2012) The emergence of sustainable manufacturing practices. Prod Plann Control 23(5):354–376. https://doi.org/10.1080/09537287.2011.555425
13. Dubey R, Bag S, Ali SS (2014) Green supply chain practices and its impact on organisational performance: an insight from Indian rubber industry. Int J Logist Syst Manag 19(1):20–42
14. Dubey R, Gunasekaran A, Childe SJ, Blome C, Papadopoulos T (2019) Big data and predictive analytics and manufacturing performance: integrating institutional theory, resource-based view and big data culture. Br J Manag 30:341–361. https://doi.org/10.1111/1467-8551.12355
15. Dubey R, Gunasekaran A, Childe SJ, Bryde DJ, Giannakis M, Foropon C, Roubaud D, Hazen BT (2020) Big data analytics and artificial intelligence pathway to operational performance under the effects of entrepreneurial orientation and environmental dynamism: A study of

manufacturing organisations. Int J Prod Econ 226:107599. https://doi.org/10.1016/j.ijpe.2019.107599
16. Duong LN, Al-Fadhli M, Jagtap S, Bader F, Martindale W, Swainson M, Paoli A (2020) A review of robotics and autonomous systems in the food industry: From the supply chains perspective. Trends Food Sci Technol 106:355–364
17. Ellis S (2018) The path to a thinking supply chain, John Santagate, IDC Technology Spotlight. https://www.ibm.com/account/reg/signup?formid=urx-32367. Accessed 29 Dec 2022
18. EPA report (2020) www.epa.gov/sustainability/sustainable-manufacturing. Accessed 28 Dec 2022
19. Esa-automation.com (2021) The Impact of Robotics in the Supply Chain. Available at: https://www.esa-automation.com/en/the-impact-of-robotics-in-the-supply-chain/#:~:text=Robotics%20in%20the%20supply%20chain%20is%20a%20key,and%20much%20more.%20Robotics%20of%20the%20future%2C%20today. Accessed 28 Dec 2022
20. Ganesh AD, Kalpana P (2022) Future of artificial intelligence and its influence on supply chain risk management–A systematic review. Comput Ind Eng 169(108206)
21. Garetti M, Taisch M (2012) Sustainable manufacturing: trends and research challenges. Prod Plann Control 23(2–3):83–104
22. Hartmann J (2021) Toward a more complete theory of sustainable supply chain management: the role of media attention. Supply Chain Manag 26(4):532–547. https://doi.org/10.1108/SCM-01-2020-0043
23. Hartmann J, Moeller S (2014) Chain liability in multitier supply chains? Responsibility attributions for unsustainable supplier behavior. J Oper Manag 32(5):281–294
24. Ivanov D (2021) Lean resilience: AURA (Active usage of resilience assets) framework for post-covid-19 supply chain management. Int J Logist Manag 33(4):1196–1217. https://doi.org/10.1108/IJLM-11-2020-0448
25. Kamble SS, Gunasekaran A, Gawankar SA (2020) Achieving sustainable performance in a data-driven agriculture supply chain: a review for research and applications. Int J Prod Econ 219:179–194
26. Klumpp M, Zijm H (2019) Logistics innovation and social sustainability: How to prevent an artificial divide in human–computer interaction. J Bus Logist 40(3):265–278
27. Luther D (2020) Supply chain sustainability: why it is important & best practices. Oracle Netsuite Publication. https://www.netsuite.com/portal/resource/articles/erp/supply-chain sustainability.shtml#:~:text=What%20Is%20Supply%20Chain%20Sustainability,every%20transportation%20link%20in%20between. Accessed 26 Dec 2022
28. Machado CG, Winroth MP, Ribeiro da Silva EHD (2020) Sustainable manufacturing in Industry 4.0: an emerging research agenda. Int J Prod Res 58(5):1462–1484. https://doi.org/10.1080/00207543.20
29. Manning L, Brewer S, Craigon PJ, Frey J, Gutierrez A, Jacobs N, Kanza S, Munday S, Sacks J, Pearson S (2022) Artificial intelligence and ethics within the food sector: developing a common language for technology adoption across the supply chain. Trends Food Sci Technol 125:33–42
30. Mitlin D (1992) Sustainable development: a guide to the literature. Environ Urban 4(1):111–124. https://doi.org/10.1177/095624789200400112
31. Mkrttchian V (2021) Artificial and natural intelligence techniques as IoP-and IoT-based technologies for sustainable farming and smart agriculture. In: Artificial intelligence and IoT-based technologies for sustainable farming and smart agriculture. IGI Global, pp 40–53
32. Nayal K, Kumar S, Raut RD, Queiroz MM, Priyadarshinee P, Narkhede BE (2022) Supply chain firm performance in circular economy and digital era to achieve sustainable development goals. Bus Strateg Environ 31(3):1058–1073
33. Naz F, Agrawal R, Kumar A, Gunasekaran A, Majumdar A, Luthra S (2022) Reviewing the applications of artificial intelligence in sustainable supply chains: exploring research propositions for future directions. Bus Strateg Environ 31(5):2400–2423
34. Olan F, Liu S, Suklan J, Jayawickrama U, Arakpogun EO (2022) The role of artificial intelligence networks in sustainable supply chain finance for food and drink industry. Int J Prod Res 60(14):4418–4433

35. Pournader M, Ghaderi H, Hassanzadegan A, Fahimnia B (2021) Artificial intelligence applications in supply chain management. Int J Prod Econ 241:108250
36. Riahi Y, Saikouk T, Gunasekara A, Badraoui I (2021) Artificial intelligence applications in supply chain: a descriptive bibliometric analysis and future research directions. Expert Syst Appl 173:114702
37. Rogers PP, Jalal KF, Boyd JA (2007) An introduction to sustainable development, 1st edn. Routledge London. https://doi.org/10.4324/9781849770477
38. Rosen MA, Kishawy HA (2012) Sustainable manufacturing and design: Concepts, practices and needs. Sustainability 4(2):154–174
39. Sanders NR, Boone T, Ganeshan R, Wood JD (2019) Sustainable supply chains in the age of AI and digitization: research challenges and opportunities. J Bus Logist 40(3):229–240
40. Shankar KM, Kannan D, Kumar PU (2017) Analyzing sustainable manufacturing practices–a case study in Indian context. J Clean Prod 164:1332–1343
41. Shi J, Li H, An H, Guan J, Arif A (2019) Tracing carbon emissions embodied in 2012 Chinese supply chains. J Clean Prod 226:28–36
42. Singh R (2022) Artificial intelligence and machine learning in supply chain management. https://www.eazystock.com/uk/blog-uk/artificial-intelligence-and-machine-learning-in-supply-chain-management/#:~:text=Machine%20learning%20makes%20better%20use%20of%20resources&text=Automating%20these%20processes%20will%20save,areas%20of%20inefficiency%20and%20waste. Accessed 29 Dec 2022
43. Song X, Xu B, Zhao Z (2022) Can people experience romantic love for artificial intelligence? An empirical study of intelligent assistants. Inf Manage 59(2):103595
44. Toorajipour R, Sohrabpour V, Nazarpour A, Oghazi P, Fischl M (2021) Artificial intelligence in supply chain management: a systematic literature review. J Bus Res 122:502–517
45. United Nations Report (2022) World population projected to reach 9.8 billion in 2050, and 11.2 billion in 2100. https://www.un.org/en/development/desa/population/events/pdf/other/21/21June_FINAL%20PRESS%20RELEASE_WPP17.pdf. Accessed 02 Mar 2023
46. Watson I (2022) The AI journey: artificial intelligence and the supply chain. https://www.ibm.com/downloads/cas/MVOQEOAB. Accessed 29 Dec 2022

Using Blockchain for Agro-Food Traceability: A Case Study from Olive Oil Industry

Abdlerahman Habashneh, Almoayied Assayed, and Amro AlMajali

Abstract Ensuring the safety and high quality of food is becoming increasingly important. Traceability of food products along the supply chain, particularly in the olive oil industry, is highly necessary to ensure their quality and provenance. There are multiple methods for traceability in the agro-food industry, such as Near-Field Communication (NFC), Radio Frequency Identification (RFID), and molecular analysis. In the era of Industry 4.0, blockchain has emerged as an appealing approach for food traceability due to its ability to provide end-to-end traceability, from raw materials acquisition to product delivery to end customers. It also creates time-stamped, unalterable, auditable, and immutable transactions. This pilot research, documented in this chapter, aims to demonstrate the use of blockchain technology to track and ensure the compliance of olive oil in a selected olive mill in Jordan. The Decapolis Food Guard (DFG) platform was adopted to achieve the objectives of this study. The process began with receipt processing, which involved entering data on farmers and olives into the DFG application, including weight, olive type, source, and other relevant information. This was followed by entering all the data related to oil processing and implementing procedures for tamper-proofing and label registration. Tamper-proof labels were applied to the spout of each container and registered in the DFG system. These labels are protected to prevent copying onto other containers with unknown contents or provenance. The users of blockchain-based technology (DFG) to validate the provenance of olives, production processes, and handling of olive oil expressed high satisfaction with the system. The owner of the olive mill confirmed a 15–25% increase in sales after introducing this technology, as customers became much more confident about the authenticity of the oil. Additionally, the technology has improved the mill's capability to export its products to other Middle Eastern and North African (MENA) and EU countries, as the oil has become compliant with international standards. It is recommended to scale up this technology and to raise

A. Habashneh · A. AlMajali
Decapolis Company, Amman, Jordan

A. Assayed (✉)
Water, Environment and Climate Change Centre, Royal Scientific Society, Amman, Jordan
e-mail: almoayied.assayed@rss.jo

K E K Vimal et al. (eds.), *Industry 4.0 Technologies: Sustainable Manufacturing Supply Chains*, Environmental Footprints and Eco-design of Products and Processes,
https://doi.org/10.1007/978-981-99-4819-2_3

awareness of all stakeholders about the importance of this technology in improving the agro-food industry.

Keywords Traceability · Olive Oil · Blockchain · Decapolis Food Guard (DFG) · Radio Frequency Identification (RFID)

1 Introduction

Traceability refers to the capacity to track a product's journey through each stage, including production, processing, and distribution, while monitoring the origin and history of materials. This capability is crucial for preventing intentional or accidental mislabeling [1, 14]. In sectors like the food-agro industry, such as olive oil production, traceability, etc., plays a vital role due to the involvement of multiple stakeholders from production to consumption [4]. Globally, approximately 3 million tons of olive oil are produced each year from around 750 million productive olive trees, with the majority concentrated in the Mediterranean region [6]. Various types of olive oil exist, including extra virgin oil, virgin olive oil, ordinary virgin olive oil, refined olive oil, and olive pomace. Olive oil holds significant cultural and dietary importance in the Mediterranean region and contributes significantly to the socio-economic well-being of local communities, especially in Jordan. However, due to its relatively high cost, olive oil is susceptible to adulteration and mislabeling, underscoring the need for reliable methods to verify its quality.

There are several methods available to verify the quality of olive oil. Some of these methods involve DNA isolation, known as molecular traceability, while others rely on chemical analysis using techniques such as amplified fragment length polymorphism, proton transfer reaction mass spectrometry (PTR-MS), nuclear magnetic resonance spectroscopy (NMR), or high-performance liquid chromatography (HPLC) [9, 10]. However, these methods are laboratory-based, expensive, and not easily accessible to consumers. With the advent of digital transformation in the current era, consumers now have the ability to obtain product information through the use of smart tags, such as radio frequency identification (RFID) and near-field communication (NFC) [14]. RFID has been successfully demonstrated by Yu-Chia et al. [5] and Rahman et al. [13] for traceability of fish throughout the supply chain. Papetti et al. [11] proposed the use of RFID for Italian cheese traceability and found it to be effective. However, RFID is a centralized process, and not all stakeholders have direct access to the data.

More recently, near-field communication (NFC) has been introduced in oil bottles. NFC serves as a link to a website without integration with a food traceability system, allowing consumers to access product details using smartphones and NFC technology. The main difference between NFC and RFID lies in the fact that NFC has been widely integrated into mobile devices such as smartphones and tablets. This has enabled the development of applications specifically designed for end-users. Figure 1 illustrates the different tracing methods available.

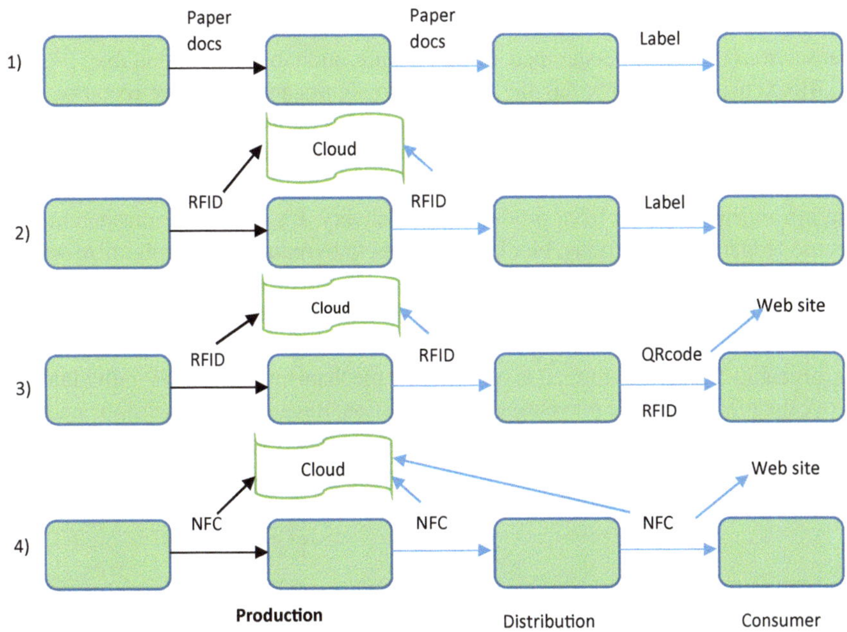

Fig. 1 Different traceability systems adapted from Pigini and Conti [12]

However, most of the traceability systems for food nowadays rely on centralized infrastructure and require third-party intermediary that result in a lack of transparency and security threats. They also hinder external stakeholders from checking for regulatory compliance.

In conventional systems, this process is costly, not feasible, or even unattainable. Business systems typically involve diverse data types originating from various sources. Auditing such inputs by "replaying" the sequence can present significant technical difficulties. Additionally, auditing may necessitate reliable knowledge and confirmation of operator identity, which can be compromised or unreliable in systems involving multiple actors. By incorporating robust and flexible data systems that ensure complete authenticity for every interaction, it becomes possible to establish resilient frameworks that are resistant to coercion and human-related issues at their core.

Presently, a novel technology known as blockchain introduces a new perspective on food traceability. The blockchain is a recent advancement in the technological realm that utilizes global peer-to-peer consensus networks to establish an open platform capable of offering neutrality, transparency, reliability, and security. Originally conceived as a solution for managing the shared accounting ledger of the Bitcoin digital currency, the underlying mechanism of blockchain has evolved. This technology enables comprehensive traceability throughout the entire flow, from the acquisition of raw materials to the delivery of products to customers [2]. It generates transactions that are time-stamped, unchangeable, auditable, and immutable (Fig. 2).

Blockchain technology has the potential to significantly improve recourse efficiency in a variety of industries. By creating a decentralized and immutable ledger of transactions, blockchain can reduce costs, increase transparency, and streamline processes. In industries such as food and agriculture, blockchain can be used to track the movement of goods from production to delivery. By creating a transparent and secure ledger of transactions, blockchain can help to reduce fraud, minimize waste, and increase efficiency. It can also enable companies to track the origin of their products, allowing them to ensure that they are ethically sourced and produced.

The blockchain only accepts authenticated information, which takes the form of an unforgeable digital signature, a cryptographic mechanism that allows individuals to prove their identity while preventing impersonation. Regardless of one's job or access capabilities, interaction with the blockchain is only possible by providing the unique digital access code associated with the user's ownership. In other words, unless

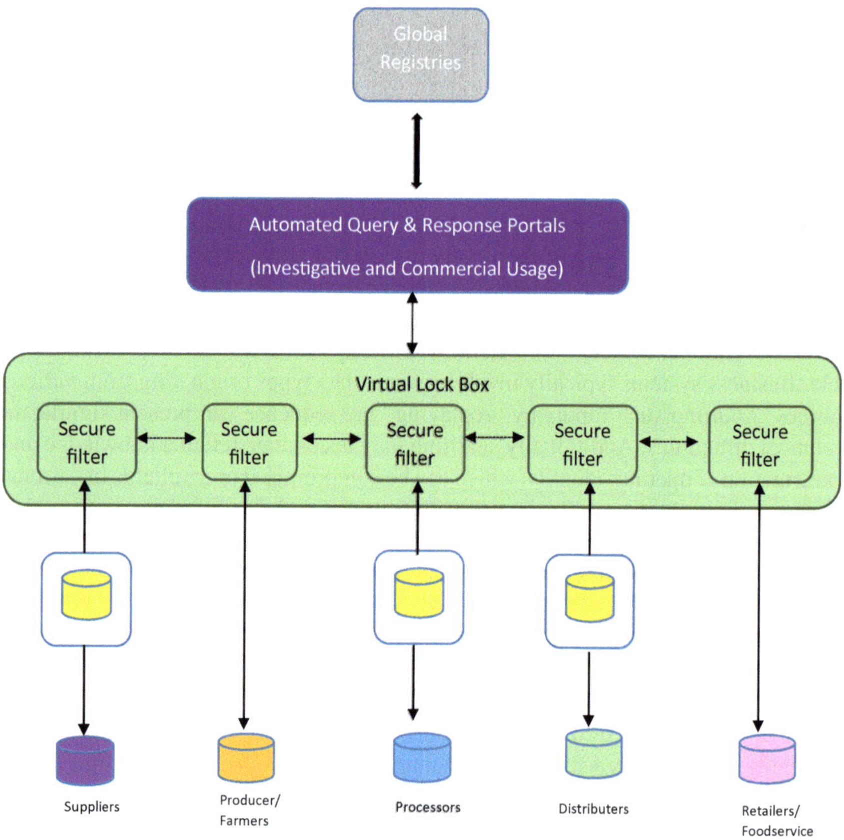

Fig. 2 Architecture in food traceability systems as adapted from Burke [2]

the ownership of the account is cryptographically proven, no one else can modify it. This approach eliminates the need for elevated privilege levels and significantly reduces the security risk associated with operators and IT administrators, who are often considered the weakest link in security systems.

The purpose of this paper is to document using blockchain technology in olive oil production in Jordan in order to validate the provenance of olives, production processes, and handling of olive oil. The rationale behind this idea was to secure the 16-L container that stores the finished oil at the olive mills vicinity, in a way that allows easy detection of tampering. Most of the finished oil in Jordan is stored in a 16-L metal container that has a single spout on it. This spout is the only place where oil can be put into and taken from the container. This spout can be secured in a way that makes it clear that the container has been opened by using some marker that will be altered if the container has been opened. Some information about the oil can be also identified through this marker. This idea has been first tested in Jordan by a start-up named "Decapolis", where blockchain traceability guard was developed, by which each batch of oil production can easily be referred to.

2 Methodology

Decapolis Food Guard (DFG) platform was used in this pilot study (https://www.decapolis.io/). The Decapolis Food Guard (DFG) platform uses blockchain technology to store tamper-proof data and smart contracts for quality assessments and procedures to be executed during all the stages of farming, production, and delivery of the product. Each stakeholder applies certified processes targeted to each stage under the guidelines and tools provided by Decapolis. Decapolis platform implements contractual relationships with each stakeholder to abide by the quality processes agreed upon by contract to be executed.

To showcase the capabilities of the Decapolis Food Guard system, the research team chose a specific olive mill situated in the southern region of Jordan. This particular mill has a seasonal production output of approximately 170 oil containers. The chosen olive mill is owned by a prominent figure who holds the position of Head of Syndicate of Olive Mills Owners in Jordan, making them an influential leader within the industry. The underlying assumption behind selecting this mill was that if the pilot implementation proves successful, it would likely encourage other mills to adopt the system as well.

The tamper-proofing approach that was used in the selected olive mill is summarized in Fig. 3. The process started by receipt processing which includes entering farmers and olive data into the DFG application (weight, olive type, source, and others). This is followed by entering all data related to the oil processing and then implementing tamper-proofing and label registration procedures, which includes affix tamper-proof labels over spout of each container that are registered at the DFG system (Fig. 4). All labels are associated with the traceability batch number during the registration. A website/portal is provided with a standard URL encoded in the

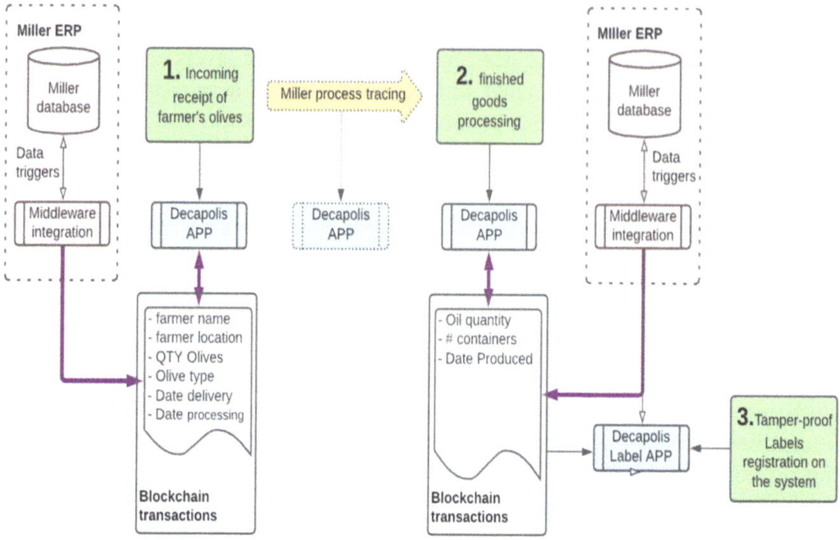

Fig. 3 Decapolis tamper-proofing steps for olive oil containers

label. Once the label is scanned, two sets of information can be displayed,off i.e., customers and miller production information. The miller is able to login in with credentials for private access to detailed production information, while the consumer needs to provide the telephone number that is verified by the OTP process to access the consumer information screen (Fig. 5).

The miller has the authority to determine which information from the factory should be recorded and which standards and regulations should be implemented for validation purposes. With the Decapolis Food Guard (DFG) system, all relevant quality and monitoring organizations, including the Jordan Food and Drug Association (JFDA), have access to comprehensive traceability data provided by the millers. The marker placed on each container cannot be replicated, ensuring that it cannot be transferred to other containers with unknown contents or uncertain origins. As a result, the DFG system ensures that the container remains tamper-proof, and when properly sealed and secured, guarantees that the contents comply with the required standards for olive oil.

3 Results and Discussion

The pilot research conducted using the Decapolis Food Guard (DFG) system examined the application of blockchain technology to validate the origin of olives, production processes, and handling of olive oil. The targeted olive mill expressed a high level of satisfaction with the system. They reported a significant increase in sales, ranging from 15 to 25%, after implementing this technology, as customers gained

Fig. 4 The oil container labeled with DFG fixed on the spout

greater confidence in the authenticity of the oil. The technology also enhanced the mill's ability to export its products to other regions such as the Middle East, North Africa, and European Union countries, as the oil met international standards.

Consumers also expressed high satisfaction with the system, highlighting its simplicity in retrieving all necessary traceability data through a quick scan using their mobile phones (Fig. 6). These findings align with existing literature that documents the sales growth and export potential facilitated by blockchain traceability systems [2, 4, 12, 14].

However, it is crucial to raise awareness among customers and other stakeholders about the importance of such technology. The lack of collaboration among actors in the supply chain, along with limited digital skills, were identified as the main challenges for the full integration of the DFG system. These challenges are consistent with the observations made by Kamilaris et al. [7] and Kumar et al. [8].

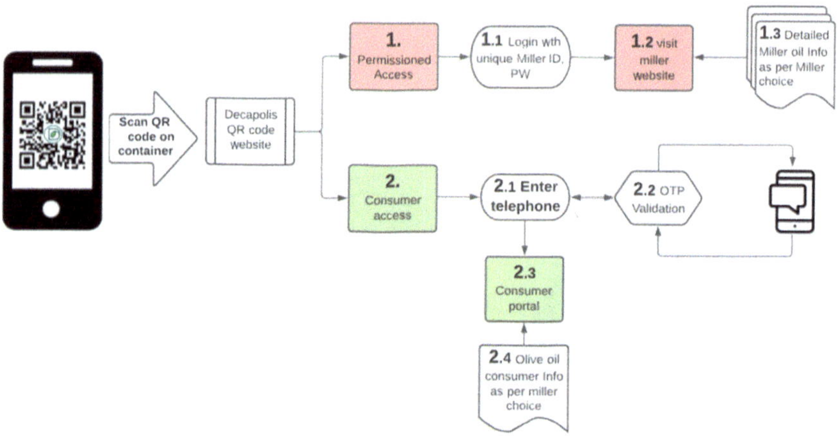

Fig. 5 The two sets of information that are displayed off when the QR of DFG is scanned

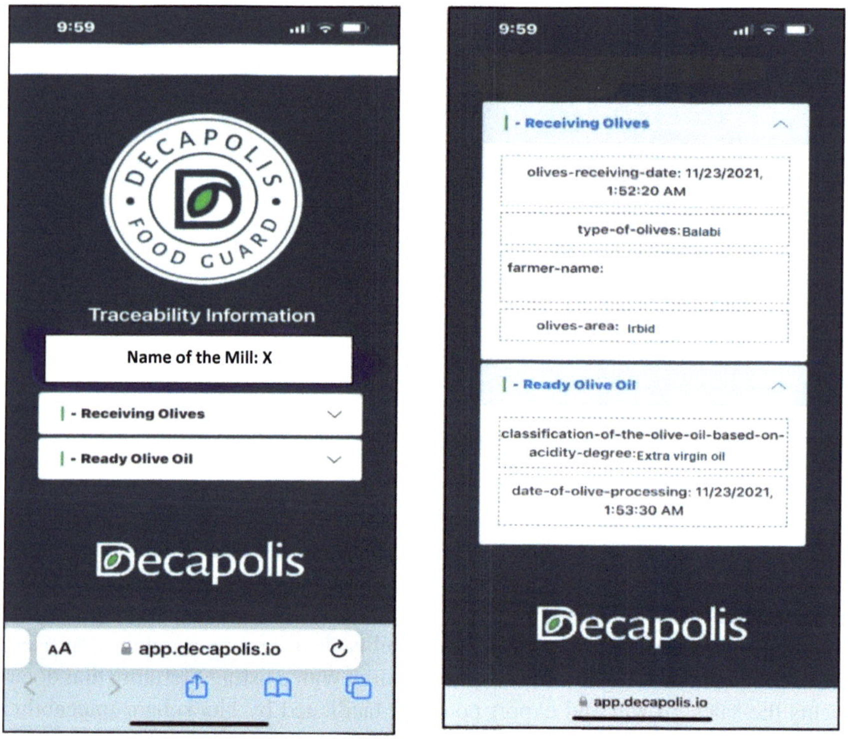

Fig. 6 A screenshot of the DFG system that appears upon scanning the QR code fixed on the olive container

Table 1 The main differences between DFG that was used in this study and RFID

Decapolis food guard (DFG)	RFID
Simple labels uniquely identify contents of a container with a QR code	RFID tags can be placed on the container to identify contents
QR code encodes address of website with unlimited content	Simple, cheap passive RFID tags properties, limited data storage, ~128 bits, 16 bytes, not enough to encode a website address
QR codes will be preprinted on an inexpensive roll dispenser	Tags will be programmed with serial number data at the tag manufacturer
No special printing hardware required	Expensive printer required to program the RFID tag
Labels are branded with Miller logo	Branded RFID Tags are more expensive
No special reader hardware needed. Only a mobile phone	Tags require reader hardware to be purchased by the Miller and available for all places where the tags need to be read; mobile readers, cost ~$500
QR code scanning distance 500 cm, sufficient for convenient access	RFID tags for scanning larger quantities of product at a longer distance requires more expensive tags
Tamper-proofing based on QR code scanning software. Access to data requires valid telephone number	Not possible to build-in tamper-proofing using RFID without expensive self-destruct capability
Scanning software resolves Duplication problem	The Duplication problem was not resolved by RFID tags: many copies can be made and be valid through RFID tag and thus applying to many containers of unknown origin

Through this pilot study, the DFG system has proved to be much more reliable comparing to RFID (Table 1). As a result of this pilot study, 10 of the biggest olive oil mills in Jordan out of the 135 mills showed high interest to subscribe in DFG so as to create trust in their products and increase sales while guaranteeing safety and quality.

4 Conclusions

The following conclusions can be withdrawn from this article:

1. Blockchain-based technology (DFG) has been proven to be a promising technology to prevent olive oil fraud and mislabeling. However, it can be extended to cover other products that are subject to alteration such as honey. The proof of concept shown in this pilot research was successful as the demand for DFG-labeled containers was high due to the blockchain technology which creates an unbroken chain of tamper-proof records.

2. Blockchain-based technology (DFG) is more reliable and robust when compared with other digital traceability systems such as RFID.
3. Decapolis Food Guard (DFG) has improved the value of olive oil containers where the consumers were willing to pay 30% more to buy DFG labeled containers.
4. Awareness and digital education of local farmers and all actors along the supply chain are quite crucial to achieve full integration of digital food traceability in the agro-food sector.
5. Overall, blockchain has the potential to significantly improve recourse efficiency in a variety of industries. By creating a decentralized and immutable ledger of transactions, blockchain can reduce costs, increase transparency, and streamline processes, leading to greater efficiency and sustainability.

References

1. Abenavoli LM, Cuzzupoli F, Chiaravalloti V, Proto AR (2016) Traceability system of olive oil: a case study based on the performance of a new software cloud. Agron Res 14(4):1247–1256
2. Burke T, McEntire J, Kennedy AW (eds) (2019) Food traceability, food microbiology and food safety. Springer Nature Switzerland AG. https://doi.org/10.1007/978-3-030-10902-8_1
3. Conti M (2017) EVO-NFC: extra virgin olive oil traceability using NFC suitable for small-medium farms. IEEEAccess (10)
4. Feng H, Wang X, Duan Y, Jian Z (2020) Applying blockchain technology to improve agri-food traceability: A review of development methods, benefits and challenges. J Clean Prod:260
5. Hsu YC, Chen A-P, Wang CH (2008) A RFID-enabled traceability system for the supply chain of live fish. In: IEEE International conference on automation and logistics, pp 81–86. https://doi.org/10.1109/ICAL.2008.4636124
6. https://www.internationaloliveoil.org/wp-content/uploads/2021/12/IOC-Olive-Oil-Dashboard-1.html#consumption-1. Accessed 31 Dec 2022
7. Kamilaris A, Fonts A, Prenafeta-Boldú FX (2019) The rise of blockchain technology in agriculture and food supply chains. Trends Food Sci Technol. https://doi.org/10.2139/ssrn.2849251
8. Kumar NM, Mallick PK (2018) Blockchain technology for security issues and challenges in IoT. Proc Comput Sci 132:1815e1823. https://doi.org/10.1016/j.procs.2018.05.140
9. Luykx D, Van Ruth S (2008) An overview of analytical methods for determining the geographical origin of food products. Food Chem 107(2):897–911. https://doi.org/10.1016/j.foodchem.2007.09.038
10. Luykx DM, Peters RJ, van Ruth SM, Bouwmeester H (2008) A review of analytical methods for the identification and characterization of nano delivery systems in food. Agric Food Chem 56(18):8231–47. https://doi.org/10.1021/jf8013926. Epub 2008 Aug 30. PMID: 18759445
11. Papetti P, Costa C, Antonucci F, Figorilli S (2012) A RFID web-based infotracing system for artisanal Italian cheese quality traceability. Food Control 27(1):234–241. https://doi.org/10.1016/j.foodcont.2012.03.025
12. Pigini D, Massimo Conti N (2017) FC-based traceability in the food chain. Sustainability 9:1910. https://doi.org/10.3390/su9101910, MDPI
13. Rahman LF, Alam L, Marufuzzaman M, Sumaila UR (2021) Traceability of sustainability and safety in fishery supply chain management systems using radio frequency identification technology. Foods 10:2265. https://doi.org/10.3390/foods10102265

14. Violino S, Pallottino F, Sperandio G, Figorilli S, Ortenzi L, Tocci F, Vasta S, Imperi G, Costa G (2020) A full technological traceability system for extra virgin olive oil. Foods 9:624. https://doi.org/10.3390/foods9050624

Logistics 4.0 for Sustainable Manufacturing Supply Chain

L. Aravindh Kumaran, M. Ramasubramaniam, and K. Sivakumar

Abstract Logistics 4.0 concepts have been made in recent years have close collaboration with Industry 4.0. Customization of products on large scale and developments in sustainability result in higher complications and requirements on logistics systems. To manage this complexity, a mechanism for handling the manufacturing supply chain to become automatic and sustainable is required. Sustainability and digitalization are transverse themes intersecting all parts of the manufacturing supply chain. While Industry 4.0 put forward modification in manufacturing in an astonishing way, Logistic 4.0 encourage the metamorphosis in an organization starting from procurement, manufacturing, distribution, warehousing, selling, and delivery of products to the customers to become supply chain more sustainable because where all the system depends on logistics. This chapter attempts to connect the link between the concept of sustainable manufacturing and Logistic 4.0 and make certain that the use of advanced technologies of Logistics 4.0 along with the operation principles can create value in all dimensions of sustainability. Also, the existent research work related to sustainable manufacturing and Logistics 4.0 are summarized along with conceptual framework developed by integrating the technologies and principles of Industry 4.0 with sustainable outcomes with a triple bottom line viewpoint. The human interface in Logistics 4.0 for accomplishing sustainable supply chain with the use of technologies is well defined. Also, adoption of Logistics 4.0 propelled by smart technologies will provide new services-products to the customers and will encourage the closed-loop life cycles in supply chain.

Keywords Logistics 4.0 · Sustainable manufacturing · Sustainable logistics · Logistics decentralization · Logistics virtualization · Human–machine interface

L. A. Kumaran · M. Ramasubramaniam · K. Sivakumar (✉)
Loyola Institute of Business Administration, Chennai, TN, India
e-mail: sivars2812@gmail.com

K E K Vimal et al. (eds.), *Industry 4.0 Technologies: Sustainable Manufacturing Supply Chains*, Environmental Footprints and Eco-design of Products and Processes,
https://doi.org/10.1007/978-981-99-4819-2_4

1 Introduction

Customer expectations and supply of the physical products are not matching in supply chain which makes manufacturers little disturbance in planning their manufacturing [1], which intrudes on the functions of logistics. To become a profitable part of the global economy company must concentrate on logistics to bring improvements to become competitive advantage. Now, it is essential for the manufacturing companies should focus on logistics as one of the major areas to face the competition in global markets. Global logistics market value reached US$ 4.92 Trillion in 2021 and expecting to reach US$ 6.75 Trillion in 2027. It indicates the growth of the logistics sector is tremendous after COVID lockdown restrictions and demands a major revolution in the logistics sector to face the growth which has already begun as part of the fourth industrial revolution. This brings transformation in the manufacturing value chain and logistics networks. Transformation supports widespread connections which allow the manufacturing, warehousing, and logistics equipment systems to communicate and exchange information, encouraging autonomous movements to control one another. This system promises transparency in information and management from supplier to customer with cost-effective material movement, this system is Logistics 4.0 [2, 3]. This transparent system aims to help the manufacturing firms forecasting to delivery of the product, part of the system depends on logistics. It is possible by the logistics to coordinate with manufacturing system only when digitalization of logistics activities to ensure a faster, flexible, and transparent system. Digitalization enables cooperation, connectivity, and adaptiveness in logistics activities, as it can react to the changes in the customer and market requirements [4].

Logistics 4.0 systems have concerned manufacturing industry to enable effective, adaptable, and individual services to attain customer expectations [5]. This system is impacted by many technologies, such as route optimization software, inventory management, optimized storage techniques as part of modern manufacturing systems, automated storages and retrieval using robots with tracking and real-time information sharing systems [6, 7]. Due to dynamic business situation in the contemporary world, the modern manufacturing systems are facing sustainability related measured to be embedded in the manufacturing and logistics process problems [7, 8]. In addition to the manufacturing sustainability measures, Logistics 4.0 provides solutions to sustainability challenges using the above technologies [5].

This chapter attempts to discuss the importance of Logistics 4.0 for sustainable manufacturing supply chain. The structure of the chapter is constructed as follows. Section 2 highlights the basic idea of sustainable manufacturing and detailed literature work on sustainable manufacturing supply chain. The concept, technologies, and principle of Logistics 4.0 are presented in Sect. 3. Section 4 discusses the role of Logistics 4.0 to achieve the sustainable outcomes, in particular Logistics 4.0 human–machine interface for sustainable supply chain is explained. Conclusion is described in Sect. 5.

2 Sustainable Manufacturing Supply Chain

Sustainable manufacturing aims for quality products using integrated processes and systems with minimal input resources. This manufacturing is conducive for employees, community surrounding the facility, and customers with respect to sustainability from procurement till disposal of the product. This system and process are attracting labor with less attrition, cost reduction in manufacturing using efficient resources, and improved brand image [9, 10]. To achieve this, integration of supply chain in the development of a product with various stages of supply chain has become essential, which demands a significant consumption of resources and facility [11]. Logistic 4.0 plays the central role in this integration of sustainable manufacturing system with the supply chain [12]. Logistics 4.0 is capable of handling complex distribution networks within sustainable manufacturing system and supply chain through its digitalization and transparency. The objective of sustainability is distributed initially and embedded in all the stages of supply chain with the nuances of Logistic 4.0 and adaptability of supply chain, thus the emergence of sustainable manufacturing supply chain as a significant area in research [11].

Automation, waste reduction in manufacturing, process flexibility, and collaboration in design will help for sustainable manufacturing using cloud data management system [13]. This system captures real-time data for efficient resource management and develops integrated cloud manufacturing system, which part of the sustainable manufacturing system [14]. It uses IoT infrastructure, which is used to enable a control mechanism in logistics integration as part of the sustainable manufacturing system using cloud enablers [15]. This sustainable manufacturing is part of circular economy, which recover and regenerate the used products and materials [16]. This circular manufacturing approach explores new manufacturing models and related supply chains, which includes collaboration with partners in each stage of supply chain and automation [17, 18]. Robust manufacturing route is also used with cloud facility for making intelligent decision for collaboration and automation in sustainable manufacturing supply chain [19]. Sustainable manufacturing supply chain faces few barriers like sort of domain expertise and assessment of quantitative data regarding the impact on the environment due to manufacturing processes, which can be supported by cloud manufacturing [20]. In addition, this system minimizes the negative impact on ecosystem and society [21] which becomes a good competitive manufacturing system in the long run [22].

In sustainable manufacturing supply chain, this sustainability is used to address the social challenge, mainly to improve the knowledge and skill level of employees by intrinsic motivation and training. In addition, analytics tools are also helping these systems analyze the data in real time from the source for rational decision-making [23]. The real-time data from different stakeholder touchpoints in the sustainable manufacturing supply chain has a potential possibility for improving sustainability and its triple bottom line by enhancing scraped goods, usage of raw material, and carbon footprint. This will result in the expansion of value creation of products due to retrieval of resources from post-usage of goods, that will result in revenue generation opportunity [24]. In addition, the toxicity in the production process is eliminated using stainable design, lean and green manufacturing, reverse logistics, and recycling and is possible through digitalization [25, 26]. Thus, sustainable manufacturing supply chain can result in faster lean production systems, predictive maintenance, reduction in waste, energy consumption, and better stakeholder collaboration with the help of connecting production system, information system in a closed loop network [26–28].

A sustainability system in manufacturing company has created a digital dimension which interacts with other sustainability systems with casual relations and with some magnitude This enhances the inter-sustainability system commitment to sustainability [29–31]. It is due to the collective success of planning in diverse manufacturing system and measuring the system performance [32]. The transparency in supply chain of most of the manufacturing systems could provide information to enhance environmental factors, which leads to achieving supply chain sustainability [33–35]. The environmental factors need to be embedded in industrial value chain and oriented towards sustainable manufacturing supply chain [36]. Innovation is the key to enhance the value chain in manufacturing, each stage of the product life cycle should have innovative operations in an ethical and sustainable manner [37, 38] which demands investments, knowledge, and training sustainability concepts to each stakeholder [39]. To achieve this economy in developing stages should invest in operational, technology, and human resources for enhancing sustainable manufacturing supply chain [40]. Thus, the sustainable manufacturing supply chain consists of value recovery in all the stage of its process to recapture the original distinctive characteristics from its end-of-life stage [41, 42].

3 Logistics 4.0

Logistics sector is an inherent component of any supply chain that moves huge quantities of freight and passengers. The sector has an important role to play in future sustainable manufacturing supply chains for several reasons. First, the logistics network connects major cities and smaller cities and integrates manufacturers and retailers in the supply chain. The connectivity materializes through the multi-modal nature of this system. The multi-modal network enables huge quantities of shipping movement through a dense inter-connected network of train stations, roads, railways, highways, seaports, and dry ports [43, 44]. Second, flows that happen in logistics network must rely on the inter-dependencies this sector has on energy supply chains such as fossil fuel, solar, electric, and wind. Third, the sector cuts across diverse geographies within countries serving both urban and rural areas that have different demographic characteristics which impact the demand. Fourth, the advent of Logistics 4.0 technologies has significant implications in the way the supply chains function to achieve the UN sustainable development goals (SDGs). Therefore, such an important system requires a comprehensive approach in terms of Logistics 4.0 framework to integrate all aspects of logistics within supply chain. The subsequent section attempts to highlight the role and nature of Logistics 4.0 technologies in achieving a sustainable manufacturing supply chain. It also provides a Logistics 4.0 planning framework which can be used for ensuring the sustainable nature of operations in its implementation.

3.1 Technologies of Logistics 4.0

Internet of Things (IoT): A technological model that intends for linking everything and everyone, at any moment and anyplace is IoT [45]. IoT will make to achieve contemporary and pioneering services and utilizations. Logistics is one of the fields gained with the implementation of IoT [46]. IoT offers real-time data on position and control of things circumstances [47].

Cloud Computing: A computing exemplar where tasks are allocated to amalgamation of software, services and connections obtained on the network [48]. In cloud, users can retrieve the resources as required and it can cope with wide range of diversified workloads. In logistics, cloud supply chain (CSC) is utilized as per the requirements of customers for supplying end products and package services [46] and to access the actual operational data over the supply chain [47].

Big data Analytics (BDA): The rise in the quantity of data made a necessity for establishment of technologies for analyzing the business are outlined by the concept of big data analytics [49]. The concept of "5 V" is applied in BDA are volume, velocity, variety, veracity, and value to deal with the huge quantity of data [46]. BDA assists in logistics process optimization and enhance the visibility in supply chain [47].

Digital twins: It is a holographic representation of a physical resource and all its processes updated from actual data, and employs machine learning, simulation, and thinking to assist in making decisions. Digital twins push the logistics sector to improve their operational efficiency, transparency, and triple bottom line.

Augmented Reality (AR): It is a mixture of four peripheral devices such as camera, computer, pointer, and the real-time world. AR can be viewed as placement of these four diverse elements in the physical world in a three-dimensional view. Industry started to search the way to optimize the logistic process due to its impact on the total cost of the product. From the perspective of logistics, AR can assist logistics industry in shipping, warehousing, raising the order and so on [50]. Augmented reality has potential application in warehouse management, especially in operations, inventory, collection and delivery, training personnel, examining the product and assist in assembly activity [51].

Autonomous Vehicles and Material Handling System: Autonomous vehicles that can go en route in the absence of driver interference through sleuthing the flow of traffic, road conditions, and environment. It uses technologies like lidar, odometry, GPS, radar, and computer vision to detect objects along with them. In logistics, the delivery and supply of materials by existing vehicles are replaced by autonomous vehicles to reduce the pollution and dependency on fossil fuel [50].

3.2 Principles of Logistics 4.0

The Logistics 4.0 framework relies on the underlying principles which lay the groundwork for design of any Industry 4.0 system [52]. These principles are generic in nature and suit the application in any logistics setup.

Fleet interoperability: The thought of interoperability from Industry 4.0 can be disseminated to vehicles. This is the ability to exchange vehicles in a fleet that perform the same function either from self-owned fleets or even from different manufacturers and fleet owners. Such a setup gives rise to a logistics network operating in a trusted environment, allowing the vehicles to communicate with each other enabling an awareness among the Internet work at forms the basis of intelligent decision-making [53].

Logistics Decentralization: This is the ability of logistics companies including the logistics personnel and the vehicle fleet to make decisions that are relevant for the movement of goods from one location to the other [54]. Instead of using a centralized server to pass on the decisions hierarchically downwards, such a decentralized system allows the local logistic operators to make decisions that may be more efficient. This system will also pay way for a self-regulated logistic network, increasing the ability of the network to meet the growing demand of consumers across the world.

Logistics Virtualization: Logistics virtualization refers to the use of data collected from Internet of Things to build a digital copy of the real world. Data collected from sensors connected to different vehicles are sent to the digital models which use this data to develop an abstract copy of the logistics network. It can identify the different logistical assets in use along with their current geographical locations. Using such a network allows different scenarios to be captured [55]. Different scenarios of failures can be simulated and all the necessary information such as the next steps in decision-making connected to logistics can be provided by such systems to human beings. In addition, this virtualization ability can also identify the next steps in the safety provisions that need to be incorporated when a breakdown or a failure occurs in the logistics network. The cyber-physical systems enable efficient information sharing by promoting decentralized communication and decision-making with a detailed level of granularity and minimum latency. Simulations can be carried out using the real-time data allowing the logistics network tool to optimize the available resources to enable efficient freight transport of goods from one site to other.

Real-time Logistics Capability: The idea of real-time capability in a Logistics 4.0 network includes collection of real-time data, analysis, and proactive decision-making. Specifically, pro-active decision making is the ability to react to failure of a vehicle in the logistics fleet with either the substitution of the failed vehicle with an alternative vehicle, replacement of failed parts, or re-routing of parts to appropriate centers for re-distribution. All these decisions happen in real time, allowing for faster responses to changes in demand [56]. In an ideal Logistics 4.0 network, the interconnection of humans and vehicles is set to change the way in which the current logistics networks are organized. The real-time logistics capability is expected to remove the bottlenecks in a traditional logistics network, bringing much-needed flexibility to the network.

Optimal Fleet Composition: The principles laid above in this section also pave way for a significant improvement in the way the Logistics 4.0 networks are organized. The cyber-physical systems can bring in a significant change in the way logistical fleets are organized. An efficient Logistics 4.0 network is dynamic in nature with fleet compositions adjusted constantly in response to changing demand conditions [57]. The configuration and re-configuration of such a system will be closely related with the manufacturing operations enabling a smooth transition between production and logistics activity.

Logistics Service Orientation: Logistics service orientation principle is derived from flexibility that can be built in a logistics network to make time-bound decisions through the collective use of the cyber-physical systems, Internet, and other stakeholders inside the logistics network which may include even other logistics companies [58]. This enables organizations to become more agile and flexible to react to changes in the market more quickly than they used to. It also allows compilation of huge volume of data that lie in unstructured format.

4 Logistics 4.0 and Sustainable Manufacturing

Sustainable manufacturing has been a focus in the past few decades. The term sustainability includes economic, social, and environmental dimensions. Being an intelligent decision-making setup, Logistics 4.0 is expected to contribute towards sustainability of the manufacturing supply chains. This is because the Industry 4.0 framework has the potential for efficient resource allocation across manufacturing thereby conserving the utilization of natural deposits like materials, energy, products, and water. The trends in recent years point to increasingly data-driven intelligent decisions made by various organizations. These organizations are exhibiting improved coordination and collaboration characteristics that underlie the efficient resource allocation models. For instance, closed-loop supply chain research has successfully demonstrated the ability to coordinate material, information, and financial flows between different stakeholders both in the forward and the reverse loop, where logistics play a crucial role. This enables the supply chains to plan intelligently for remanufacturing, recycling, or reusing the products and components. Future interconnected supply chains can leverage the efficient use of these resources by exchanging them in an intelligent fashion.

Traditionally, logistic activities are characterized by the use of capital-intensive assets that are used for a longer period before they are scrapped or sold. The advent of Logistics 4.0 brings a coordinated decision-making where these vehicles' life can be extended through intelligent use. The vehicles can be upgraded with suitable IoT hardware and connected to analytics software to prolong the use and extend the life of these vehicles and equipment. Such a Machine-IoT integration can also address the issue of heterogeneity among the inter-connected manufacturing networks through the appropriate use of software interface. The Logistics 4.0 network can dynamically plan the routes to use its capacity to the fullest, thereby increasing utilization of such valuable assets (Fig. 1).

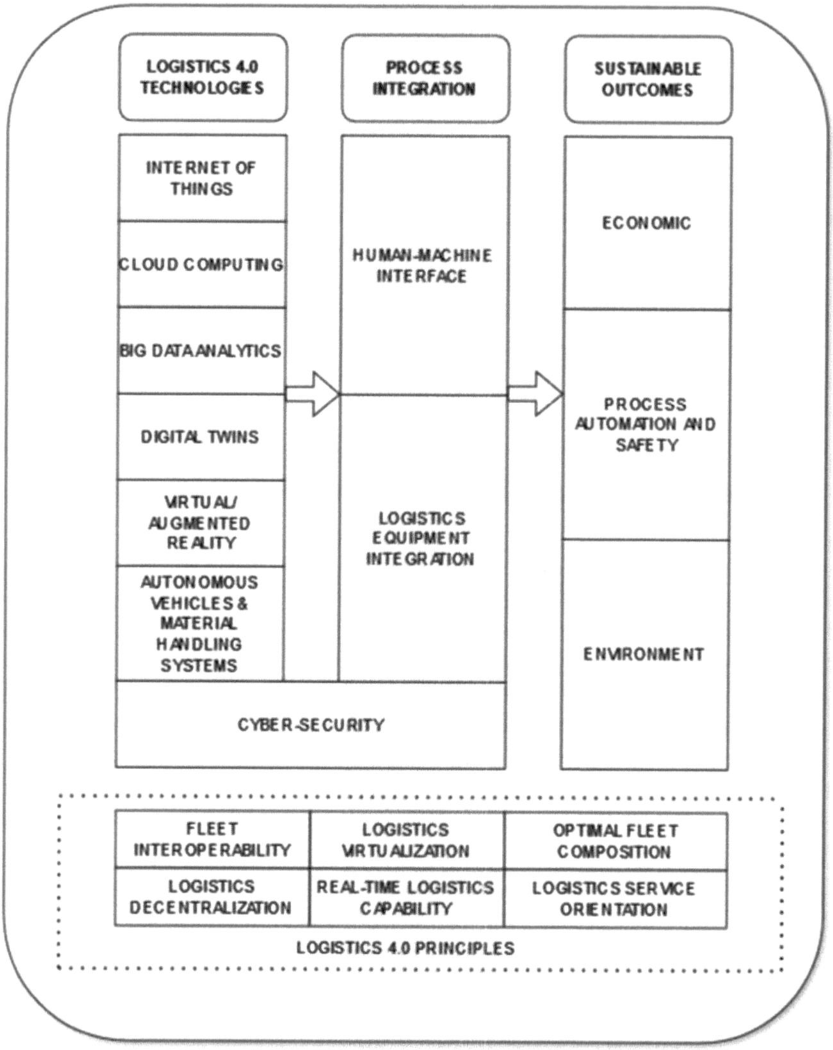

Fig. 1 Logistics 4.0 and sustainable supply chains framework

4.1 Logistics 4.0 Human–Machine Interface for Sustainable Supply Chains

The Logistics 4.0 framework throws significant opportunities for humans to redefine their relationship with machines. Unlike the doomsday prediction made for human employment, Logistics 4.0 envisages a more creative, innovative value-creation by humans compared to previous industrial revolutions. The role of humans is only going to increase by the way of creative and innovative designs of the Logistics 4.0 network. The human–machine interface in Logistics 4.0 context can follow a two-pronged approach. The first one is humans assisting machines to train in performing the required tasks. Humans are expected to train, explain, and sustain the machine operations. For instance, training the truck to operate in a complex-driving scenario. Therefore, training machines is a crucial phase where humans are expected to periodically assist the machines in making the right decisions. In the context of autonomous trucks, this could be training the truck to avoid human or other obstacles that come along the way. Vehicles also need to be trained for empathy. Although autonomous, trucks require intervention of humans in a difficult situation. This implies the presence of humans in the vehicles for a prolonged time. Autonomous trucks can be empathetic towards humans in such situations for which they need to be trained. The Explanation phase is where the people involved in the human–machine interface are expected to explain the process of decision-making done by the machines. This is especially crucial in the context of vehicle accidents and the following insurance claims. The role of explainers therefore becomes crucial. In the third phase of Sustaining, the humans involved in the design of such systems are expected to monitor the progress of machines for any required course correction mechanisms. This is extremely important to ensure the relevance and time-bound nature of machine decisions.

In the second approach, machines are expected to assist humans in three different ways. The first phase is the amplification phase, where machines can be a creative assist for decision-making. For instance, in the autonomous driving scenario, the vehicle can show dynamic routes to humans foreseeing a traffic jam thereby enabling humans to plan better. They can also become creative in the vehicle design phase, where thousands of vehicle designs can be seamlessly made in a matter of seconds. The second phase of interaction will be interesting. In this phase, the machines can interact with humans in several ways. For instance, logistics service delivery information can be stored for access by natural language assistants built-in the vehicles. This enables customers to efficiently interact with the Logistics 4.0 network. All such queries can be stored for future use by the system. The third phase is an innovative phase where machines embody humans. Intelligent agents can be embodied in autonomous vehicles with sophisticated sensors and actuators to recognize the flow of materials within a shop floor and during customer delivery. The autonomous material handling systems are expected to carry out complementary tasks alongside humans with minimal attention to safety aspects.

5 Conclusions

This chapter contributes to Logistics 4.0 for sustainable manufacturing supply chain through utilization of advanced technologies which can afford Logistics 4.0 to have beneficial effects on the dimension of sustainability. Furthermore, the concept of Logistics 4.0 and sustainable manufacturing supply chain are well defined with the thorough literature along with the links between Logistics 4.0 and sustainable manufacturing supply chain. In addition, the implementation of Logistics 4.0 could aid the manufacturing industry supply chain to become sustainable from the design to delivery of products to consumers through adopting the 4.0 principles and utilizing the emerging technologies. This was also clearly explained with human–machine interface in Logistics 4.0 for attaining sustainable supply chain. In future, the focus can be shifted to opportunities linking other aspects of sustainable manufacturing with Logistics 4.0 to make industry to accomplish maximum benefits.

References

1. Yin Y, Stecke KE, Li D (2018) The evolution of production systems from Industry 2.0 through Industry 4.0. Int J Prod Res 56(1–2):848–861
2. Schuh G, Deindl M (2013) Systematisation of smart objects in production and logistics applications. In: Smart SysTech 2013; European conference on smart objects, systems and technologies. VDE, Erlangen/Nuremberg, Germany, pp 1–9
3. Jung JU, Kim HS (2015) Big data governance for smart logistics: a value-added perspective. In: Proceedings of the 15th international conference internet of things, smart spaces, and next generation networks and systems. Springer, St. Petersburg, Russia, pp 95–103
4. Kayikci Y (2018) Sustainability impact of digitization in logistics. Proc Manuf 21:782–789
5. Winkelhaus S, Grosse EH (2020) Logistics 4.0: a systematic review towards a new logistics system. Int J Prod Res 58(1):18–43
6. Barreto L, Amaral A, Pereira T (2017) Industry 4.0 implications in logistics: an overview. Proc Manuf 13:1245–1252
7. Strandhagen JO, Vallandingham LR, Fragapane G, Strandhagen JW, Stangeland ABH, Sharma N (2017) Logistics 4.0 and emerging sustainable business models. Adv Manuf 5:359–369
8. Witkowski K (2017) Internet of things, big data, industry 4.0–innovative solutions in logistics and supply chains management. Proc Eng 182:763–769
9. Gunasekaran A, Spalanzani A (2012) Sustainability of manufacturing and services: investigations for research and applications. Int J Prod Econ 140(1):35–47
10. Bonvoisin J, Stark R, Seliger G (2017) Field of research in sustainable manufacturing. Springer International Publishing, pp 3–20
11. Hofmann E, Rüsch M (2017) Industry 4.0 and the current status as well as future prospects on logistics. Comput Ind 89:23–34
12. Delfmann W, Ten Hompel M, Kersten W, Schmidt T, Stölzle W (2018) Logistics as a science: central research questions in the era of the fourth industrial revolution. Logist Res 11(9):1–13
13. Fisher O, Watson N, Porcu L, Bacon D, Rigley M, Gomes RL (2018) Cloud manufacturing as a sustainable process manufacturing route. J Manuf Syst 47:53–68
14. Qu T, Lei SP, Wang AA, Nie DX, Chen X, Huang GQ (2016) IoT-based real-time production logistics synchronization system under smart cloud manufacturing. Int J Adv Manuf Technol 84(1):147–164

15. Zhou L, Zhang L, Fang Y (2020) Logistics service scheduling with manufacturing provider selection in cloud manufacturing. Rob Comput Integr Manuf 65:101914
16. Waste & Resources Action Programme (WRAP) (2022) WRAP and the circular economy. http://www.wrap.org.uk/about-us/about/wrap-and-circular-economy. Accessed 13 May 2022
17. Da Silveira G, Borenstein D, Fogliatto FS (2001) Mass customization: literature review and research directions. Int J Prod Econ 72(1):1–13
18. Mahesh M, Ong SK, Nee AYC, Fuh JYH, Zhang YF (2007) Towards a generic distributed and collaborative digital manufacturing. Rob Comput-Integr Manuf 23(3):267–275
19. Li BH, Zhang L, Wang SL, Tao F, Cao JW, Jiang XD, Chai XD et al (2010) Cloud manufacturing: a new service-oriented networked manufacturing model. Comput Integr Manuf Syst 16(1), 1–7
20. Bey N, Hauschild MZ, McAloone TC (2013) Drivers and barriers for implementation of environmental strategies in manufacturing companies. CIRP Ann 62(1):43–46
21. Bocken NM, Short SW, Rana P, Evans S (2014) A literature and practice review to develop sustainable business model archetypes. J Clean Prod 65:42–56
22. Schaltegger S, Wagner M (2011) Sustainable entrepreneurship and sustainability innovation: categories and interactions. Bus Strateg Environ 20(4):222–237
23. Zhong RY, Xu X, Klotz E, Newman ST (2017) Intelligent manufacturing in the context of industry 4.0: a review. Engineering 3(5):616–630
24. Rajput S, Singh SP (2021) Industry 4.0—challenges to implement circular economy. Benchmarking Int J 28(5):1717–1739
25. Duarte S, Cruz-Machado V (2017) Exploring linkages between lean and green supply chain and the industry 4.0. In: Xu J, Gen M, Hajiyev A, Cooke F (eds) Proceedings of the Eleventh international conference on management science and engineering management, Springer, Cham, pp 1242–1252
26. Waibel MW, Steenkamp LP, Moloko N, Oosthuizen GA (2017) Investigating the effects of smart production systems on sustainability elements. Proc Manuf 8:731–737
27. Hermann M, Pentek T, Otto B (2016) Design principles for Industrie 4.0 scenarios. In: 49th Hawaii international conference on system sciences (HICSS), Koloa, HI, USA, pp 3928–3937
28. Kiel D, Müller JM, Arnold C, Voigt KI (2017) Sustainable industrial value creation: benefits and challenges of industry 4.0. Int J Innovation Manag 21(8):1740015
29. Stark R, Grosser H, Beckmann-Dobrev B, Kind S, INPIKO Collaboration (2014) Advanced technologies in life cycle engineering. Proc CIRP 22:3–14
30. Beier G, Niehoff S, Ziems T, Xue B (2017) Sustainability aspects of a digitalized industry–a comparative study from China and Germany. Int J Precis Eng Manuf Green Technol 4(2):227–234
31. Beier G, Niehoff S, Xue B (2018) More sustainability in industry through industrial internet of things? Appl Sci 8(2):219
32. Ngai EWT, Chau DCK, Poon JKL, To CKM (2013) Energy and utility management maturity model for sustainable manufacturing process. Int J Prod Econ 146(2):453–464
33. Gimenez C, Sierra V, Rodon J (2012) Sustainable operations: their impact on the triple bottom line. Int J Prod Econ 140(1):149–159
34. Dubey R, Gunasekaran A, Childe SJ, Papadopoulos T, Luo Z, Wamba SF, Roubaud D (2019) Can big data and predictive analytics improve social and environmental sustainability? Technol Forecast Soc Chang 144:534–545
35. Ruiz-Benitez R, López C, Real JC (2019) Achieving sustainability through the lean and resilient management of the supply chain. Int J Phys Distrib Logist Manag 49(2):122–155
36. Fatimah YA, Govindan K, Murniningsih R, Setiawan A (2020) Industry 4.0 based sustainable circular economy approach for smart waste management system to achieve sustainable development goals: a case study of Indonesia. J Clean Prod 269:122263
37. Garcia-Muiña FE, González-Sánchez R, Ferrari AM, Settembre-Blundo D (2018) The paradigms of Industry 4.0 and circular economy as enabling drivers for the competitiveness of businesses and territories: the case of an Italian ceramic tiles manufacturing company. Soc Sci 7(12):255

38. Yadav G, Luthra S, Jakhar SK, Mangla SK, Rai DP (2020) A framework to overcome sustainable supply chain challenges through solution measures of industry 4.0 and circular economy: an automotive case. J Clean Prod 254:120112
39. Piyathanavong V, Garza-Reyes JA, Kumar V, Maldonado-Guzmán G, Mangla SK (2019) The adoption of operational environmental sustainability approaches in the Thai manufacturing sector. J Clean Prod 220:507–528
40. Thakur V, Mangla SK (2019) Change management for sustainability: Evaluating the role of human, operational and technological factors in leading Indian firms in home appliances sector. J Clean Prod 213:847–862
41. Chang AY, Cheng YT (2019) Analysis model of the sustainability development of manufacturing small and medium-sized enterprises in Taiwan. J Clean Prod 207:458–473
42. Pichagonakesit T, Ueasangkomsate P, Sudharatna Y (2019) A review of trends in sustainable manufacturing. In: 2019 Joint international conference on digital arts, media and technology with ECTI northern section conference on electrical, electronics, computer and telecommunications engineering, IEEE, Nan, Thailand, pp 202–205
43. Sammon JP, Caverly RJ (2007) Transportation systems: critical infrastructure and key resources sector-specific plan as input to the National Infrastructure Protection Plan. Department of Homeland Security, Washington DC, USA
44. Möller DPF, Deriyenko T, Vakilzadian H (2015) Cyber-physical vehicle tracking system: Requirements for using a radio frequency identification technique. In: 2015 IEEE International conference on electro/information technology, IEEE, Dekalb, IL, USA, pp 552–557
45. Lu Y, Papagiannidis S, Alamanos E (2018) Internet of Things: a systematic review of the business literature from the user and organisational perspectives. Technol Forecast Soc Chang 136:285–297
46. Corrêa JS, Sampaio M, Barros RDC (2020) An exploratory study on emerging technologies applied to logistics 4.0. Gestão Produção 27(3):1–25
47. Moldabekova A, Zhidebekkyzy A, Akhmetkaliyeva S, Baimukhanbetova E (2020) Advanced technologies in improving the management of logistics services: Bibliometric network analysis. Polish J Manag Stud 21(1):211–223
48. Dai H, Ge L, Zhou W (2015) A design method for supply chain traceability systems with aligned interests. Int J Prod Econ 170:14–24
49. Roßmann B, Canzaniello A, von der Gracht H, Hartmann E (2018) The future and social impact of big data analytics in supply chain management: results from a Delphi study. Technol Forecast Soc Change 130:135–149
50. Akkaya M, Kaya H (2019) Innovative and smart technologies in logistics. In: 17th International logistics and supply chain congress, Istanbul, Turkey, pp 97–105
51. Cano JA, Salazar-Arrieta F, Gómez Montoya RA, Cortés P (2021) Disruptive and conventional technologies for the support of logistics processes: a literature review. Int J Technol 12(3):448–460
52. Carvalho N, Chaim O, Cazarini E, Gerolamo M (2018) Manufacturing in the fourth industrial revolution: a positive prospect in sustainable manufacturing. Proc Manuf 21:671–678
53. Neto AAT, de Barros F, Ramos A, Moraes MMR, Santiago SB, de Souza Junior AA (2022) Logistics interoperability as a boost factor for Industry 4.0: case study of a motorcycle manufacturer. Eur J Bus Manag Res 7(2):69–78
54. Humayun M, Jhanjhi N, Hamid B, Ahmed G (2020) Emerging Smart logistics and transportation using IoT and blockchain. IEEE Internet Things Mag 3(2):58–62
55. Lukáč S Jr, Mikušová N (2019) Virtualization as a logistics support for enterprise management. Transp Logist 19(46):22–27
56. Kong XT, Zhong RY, Zhao Z, Shao S, Li M, Lin P, Chen Y, Wu W, Shen L, Yu Y, Huang GQ (2020) Cyber physical ecommerce logistics system: an implementation case in Hong Kong. Comput Ind Eng 139:106170

57. Vieira BS, Ribeiro GM, Bahiense L, Cruz R, Mendes AB, Laporte G (2021) Exact and heuristic algorithms for the fleet composition and periodic routing problem of offshore supply vessels with berth allocation decisions. Eur J Oper Res 295(3):908–923
58. Pan S, Zhong RY, Qu T (2019) Smart product-service systems in interoperable logistics: design and implementation prospects. Adv Eng Inform 42:100996

Industry 4.0 in Resource Efficient and Cleaner Production: A Case Study from the Food Sector in Jordan

Almoayied Assayed, Jehan Haddad, Husam Kilani, Rawia Abdallah, and Vikas Kumar

Abstract The concentration of CO_2 in the atmosphere has reached 400 ppm, a higher level than at any time in the history. As clearly referred at IPCC reports, human activity has been behind this extraordinary increase. Transition to a net-zero at industries is one of the crucial measures to meet the international commitments regarding climate change and thus keep global warming below the 1.5 °C limit. Resource Efficient and Cleaner Production (RECP) provides a practical example of de-coupling between economic growth from environmental degradation and thus supports achieving the net-zero transition. In that context, RECP has been applied to 12 industries in Jordan that resulted in saving of over 1.6 million Jordanian Dollars (JOD) annually, 22,181 MWh/year of energy; 63,844 m^3/year of water; 404 tons/year of raw material; and reduction of 8,086 tons/year of CO_2 emissions. Integrating Industry 4.0 tools into RECP has been shown to be very efficient as the industries have become able to monitor and validate all saving data on real-time bases and optimize using natural resources accordingly. Saving measures and performance evaluation were linked with IoT sensors and blockchain at one food industry in Jordan and found to be accurate and very helpful to assess input–output outflow within the industry boundary. Employing IoT with blockchain to facilitate collecting data related to RECP at the industry level can be a great advantage to RECP and help in monitoring carbon footprint accurately. It is recommended to scale up RECP-Industry 4.0 model to other industries and leverage this experience to expedite transition to circular economy.

Keywords Resource efficient and cleaner production (RECP) · Industry 4.0 · IoT sensors · Blockchain · Circular economy

A. Assayed (✉) · J. Haddad · H. Kilani · R. Abdallah
Water, Environment and Climate Change Centre, Royal Scientific Society, P.O Box 1438, Amman 11941, Jordan
e-mail: Almoayied.assayed@rss.jo

V. Kumar
Faculty of Business, Law and Social Sciences, Birmingham City University, Birmingham B4 7BD, UK

K E K Vimal et al. (eds.), *Industry 4.0 Technologies: Sustainable Manufacturing Supply Chains*, Environmental Footprints and Eco-design of Products and Processes,
https://doi.org/10.1007/978-981-99-4819-2_5

1 Background

The rapid population growth and improvement in life standards have led to a large volume of natural resources being extracted. Resource extraction increased from almost 27 billion tons in 1970 to 92 billion tons in 2017 [1]. In recent decades, material extraction has become more concentrated in upper-middle-income countries, while the higher-income countries maintain the highest material footprint of 27 tons per capita representing 13 times of the low-income group [2]. Over the past two decades, there has been a rapid acceleration in the extraction and processing of natural resources, which is responsible for over 90% of biodiversity loss, water stress, and approximately 50% of the impacts related to climate change [3]. Therefore, decoupling is needed to stop environmental degradation while attaining economic growth [4]. The environmental impact of the industrial sector shows a continuous increase, which has coincided with continuous change in environmental strategies moving from passive to proactive through a cleaner production strategy [5].

Industrial food processing plants are characterized by significant water, energy, and material consumption. The agri-food sector is a major consumer of energy and water, where almost 30% of the world's energy is consumed within agri-food systems [6]. The agri-food industry relies on energy for various purposes, including machinery and equipment operation, transportation, processing, refrigeration, and packaging. Energy consumption can vary across different stages of the food supply chain, from farm operations to processing plants and distribution networks. Water consumption in the agri-food industry is substantial due to various activities such as irrigation for crop production, livestock watering, and food processing. It is estimated that globally, the agri-food sector accounts for around 70% of freshwater withdrawals. This includes both direct waters use in agricultural activities and indirect water use in food processing and production. The global average water footprint for sugar crops, for example, is estimated at 200 m^3/ton, fruits 1000 m^3/ton, cereals 1600 m^3/ton, and oil crops 2400 m^3/ton [7]. Nevertheless, exact figures for water and energy consumption in the agri-food industry vary by region and specific practices.

Companies themselves play a major role in optimizing the resources being used in production. It is estimated that 40–50% of raw materials of food delivered to companies are rejected for quality issues [8]. Once a food processing plant is operational and its products are available in the market, one of the most efficient methods to mitigate environmental impacts is by implementing the Resource Efficient and Cleaner Production approach (RECP). RECP is an enterprise-level approach that focuses on resource optimization, minimizing environmental pollution, and fostering sustainable industrial development [9]. By implementing RECP, the generation of waste is reduced, and the full potential of raw materials is maximized. The RECP is basically underpinned by the concept of de-coupling of economic growth from environmental degradation and is part of the global efforts to enhance the transition to the circular economy. In essence, cleaner production is a key component of the Circular Economy, as it helps to ensure that materials and products are produced in a way that is sustainable and environmentally responsible. By integrating cleaner production practices with the Circular Economy model, companies can work toward a more sustainable and circular economic model, which can ultimately help to promote economic, social, and environmental sustainability. There are multiple methods to do the RECP such as SCORE—Sustaining competitive and responsible enterprises; TEST—Transfer of Environmentally Sound Technology and PREMA—Profitable resource-efficient management [10], each of which has different features, yet with one goal, which is minimizing the ecological footprint of the products while improving the financial performance of the industries.

2 RECP Demonstration in Jordan

The RECP was demonstrated in Jordan during the period from 2015 to 2018 within the SwitchMed project. The SwitchMed initiative was a regional project funded by the European Union (EU) and implemented by the United Nations Industrial Development Organization (UNIDO) [11]. In Jordan, the project applied the TEST method for implementing RECP in the targeted industries. TEST was organically developed by UNIDO and is underlined by the concept of Material Flow Cost Accounting (MFCA) and supported by the material and energy flow information system [12]. By embracing the TEST methodology, businesses can implement sustainable production models, leading to a range of benefits. These include enhanced productivity, lowered operational expenses, improved product quality, optimized investments, reduced costs associated with environmental compliance, minimized environmental impact, expanded opportunities to enter new market segments, resilient supply chain, and stronger relationships with stakeholders.

As part of the SwitchMed project, the TEST methodology has been successfully demonstrated in 12 companies in the food and beverage sector in Jordan during the period 2015–2018. Initial assessments (IAs) were conducted to assess each company's environmental, energy, and economic potentials as well as the management motivation. TEST is implemented on the basis of the Plan-Do-Check-Act (PDCA) cycle as shown in Fig. 1. RECP teams from external consultants and internal companies' staff (finance, production, operation, quality assurance, and maintenance) were established at each targeted industry and together demonstrated the TEST assessment. Figure 1 shows an overview of the TEST methodology that was applied to the 12 industries in Jordan.

Applying the TEST method has led to the discovery of significant savings potential in 12 food and beverage plants that participated in the program. These savings include an annual energy reduction of 22,181 MWh, water savings of 63,844 m^3 per year, 404 tons per year of raw material savings, a decrease of 8,086 tons per year in CO_2 emissions, and the avoidance of 83 tons per year of solid waste going to landfills [14]. In total, the RECP team identified 214 resource efficiency measures across the 12 demonstration companies. Out of these, 161 measures (approximately 75% of the total) were approved by company management and included in the companies' plans for action. Most of these interventions had a Pay Back Period (PBP) of less

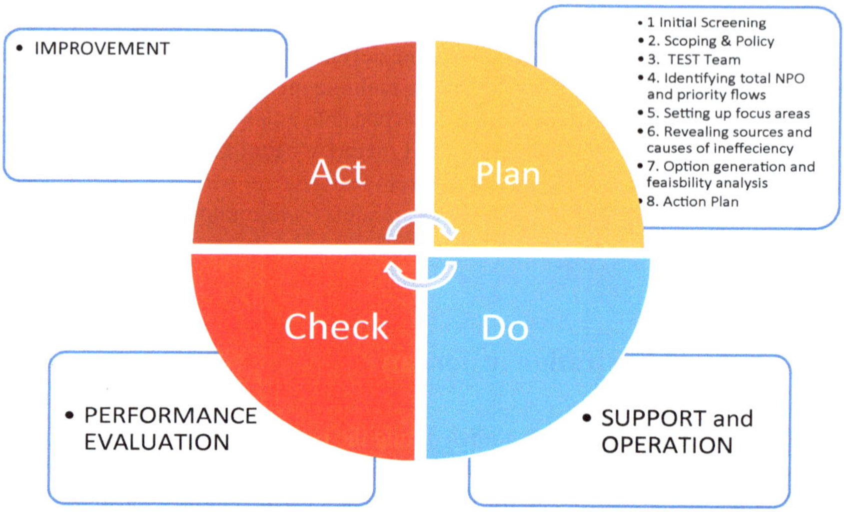

Fig. 1 Overview of the TEST approach. *Source* UNIDO TEST Tool Kit [13], available at: https://www.test-toolkit.eu/

than 6 months (48%), with an investment requirement of less than 7,000 JOD (9,300 Euros). Furthermore, around 23% of the measures were categorized as good housekeeping measures, highlighting their high profitability. Table 1 provides examples of RECP measures identified in two medium-sized food industries (with fewer than 100 employees) through the TEST approach, along with their respective payback periods.

Interestingly, most of the saving measures identified during the TEST were implemented by the companies themselves. After a year of implementation, a performance evaluation was conducted considering the benchmark and information system that was set from the very beginning to control inefficiencies that are related to resource productivity. The effective implementation of RECP options led to significant performance enhancements toward reaching the established benchmark. Figure 2 shows the performance of the water and energy-saving measures implemented in one of the Food Companies producing mainly frankfurter, frozen burgers, mortadella, breaded meats, and meatballs. The energy measures such as using LED lights, installing sub-meters to refrigerators and compressors, fixing all steam leakages, and optimizing the temperature of refrigerators, were able to reduce energy consumption by 17%. Likewise, water measures such as the reuse of treated wastewater, monitoring water supply, and installing water sub-meters, achieved an 11% drop in water consumption.

However, the regular collection of data and coordination among different departments for effective data management at the company level posed challenges in validating the RECP measures. This highlights the significance of integrating Industry 4.0 tools into the RECP concept.

Table 1 Saving opportunities at two food industries that applied TEST methodology

Measures/ saving opportunities	Main financial figures			Saving in resources per year		
	Investment (€)	Savings/ year (€)	PBP/ year	Materials and water	Energy (MWh)	Pollution reduction
Carbonated soft drinks industry						
Optimization of clean-in-place (CIP)	6,600	27,755	0.2	8,160 m^3 water	276	Total: 1,190 tons CO_2 35.9 tons solid waste 677 kg BOD_5 1,221 kg COD 40,000 m^3 recycled wastewater
Water conservation	10,133	225,060	0.1	18,210 m^3 water, 329.6 t and 40.5 m^3 raw materials	886	
Raw materials savings	2,933	20,744	0.1	12 m^3 water 4 t of sugar and 1.3 t of preform as raw materials	1	
Lighting and cooling systems	23,860	108,555	0.2	–	866	
Steam and compressed air systems	62,240	71,245	0.9	–	616	
Dairy industry						
Reducing the losses of raw materials and water	1,002,030	196,220	5.1	7.1 tons raw materials 7,505 m^3 water	205.2	Total: 209 tons CO_2 4.0 tons solid waste
Heat recovery and conservation	13,730	10,140	1.4	262 m^3 water	179.4	
Lighting and compressed air	2,030	2,610	0.8	–	24.2	
Cooling system	30,330	18,210	1.7	–	168.7	

Source SwitchMed Magazine [14], available at https://www.unido.org/sites/default/files/files/2020-01/SwitchMed%20Magazine%20-%20Jordan.pdf

3 Applying Industry 4.0 in RECP

Industry 4.0 is a term used to describe the ongoing automation and digitalization of industrial processes. It encompasses technologies such as the Internet of Things (IoT), Artificial Intelligence (AI), Big Data Analytics, and Advanced Robotics. Industry 4.0 is characterized by the ability to connect machines, devices, and people in real time, allowing for greater efficiency, productivity, and customization in manufacturing processes [15].

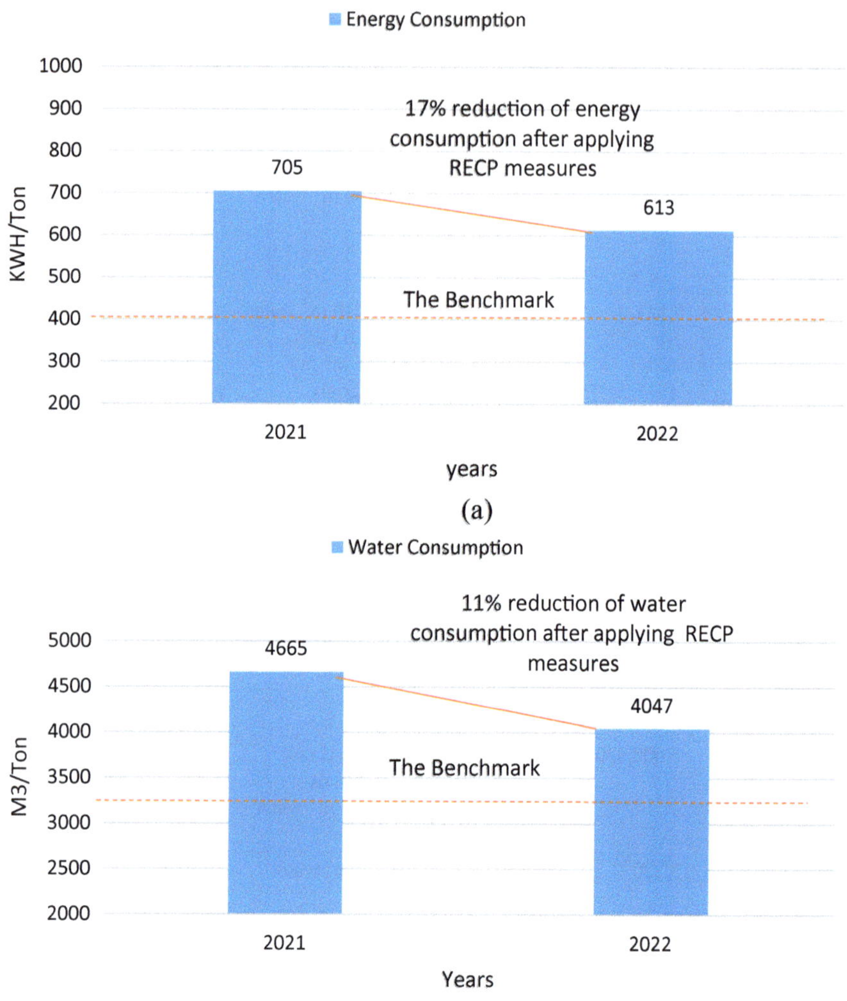

Fig. 2 **a**, **b** Reduction in energy and water consumption at one food company after applying RECP measures

As already mentioned, the agri-food industry consumes a vast amount of resources (materials, energy, and water) and generates a huge amount of food waste. It has been shown above that RECP offers significant benefits to the industries including cost saving and providing more sustainable food products. However, industries that apply RECP are struggling to digitally audit, monitor, verify, and store savings in resources, which in many cases hinder the continuous improvement of resources management at a company level. Therefore, by collecting and analyzing RECP data at the industry level in real time, better management decisions can be made to improve resource efficiency [16]. Therefore, the integration of Industry 4.0 technologies into cleaner production processes has the potential to significantly reduce the environmental impact of manufacturing. By utilizing real-time data and advanced analytics, companies can identify areas of inefficiency and waste in their production processes and optimize them for greater efficiency and reduced environmental impact. So, Industry 4.0 is seen as a key driver of sustainable development in manufacturing, and an important step toward a more circular and sustainable economy. For example, the use of advanced sensors and machine learning algorithms can help manufacturers optimize energy consumption, reduce waste, and improve the use of raw materials.

The utilization of Industry 4.0 tools in the food industry has recently gained significant attention and has been discussed in various academic literature. Noor et al. [15] examined the practical implementation of Industry 4.0 technologies in the food processing sector. They identified nine key technological advancements that are driving Industry 4.0 and explored their application areas within the food industry. These areas included intelligent manufacturing, food safety, training, and marketing. The authors emphasized the crucial role of Industry 4.0 in enhancing resource efficiency, reducing costs, expanding market access, improving customer satisfaction, implementing systematic management practices, ensuring enhanced food safety, and promoting transparency. Kumar [17] further highlights the importance of Industry 4.0 technologies in addressing the challenges faced by the food sector. In their study, Jagtap et al. [16] introduced a framework that leverages the Internet of Things (IoT) for monitoring food waste generation, as well as the consumption of energy and water in the food industry. The framework comprises three key stages: defining the necessary datasets for tracking food waste, water usage, and energy consumption; designing an IoT system to monitor and enhance resource efficiency in waste generation and consumption; and developing a decision support tool to model and design strategies for managing food waste, water usage, and energy consumption. The research paper establishes a conceptual basis for employing IoT-based technologies to monitor waste generation, water usage, and energy consumption on an industrial scale, with the aim of providing real-time information to factory management through intuitive dashboards.

The application of blockchain technology for food traceability has also received considerable attention in the existing literature. In their study, Prince et al. [18] proposed a hybrid model that combines recurrent neural networks (RNN) algorithms with IoT and blockchain technologies to forecast the supply and demand of food. The model utilizes long short-term memory (LSTM) and gated recurrent units (GRU) as prediction models while employing the Genetic Algorithm (GA) optimization to optimize the hybrid model's parameters. The research offers supply chain practitioners the opportunity to leverage cutting-edge technologies and develop policies based on the predictions generated by advanced deep learning (ADL) techniques.

Furthermore, Kayikci et al. [19] and Casino et al. [20] extensively explored the potential implementation opportunities and obstacles associated with adopting blockchain technology in the food supply chain, considering the aspects of people, process, performance, and technology. The authors highlighted that blockchain technology has the capability to address trust-related concerns within the supply chain by facilitating secure data storage with inherent privacy and management features. Kasten [21] focused on the application of blockchain technology to ensure that the results of milk analysis remain unaltered and tamper-proof, thereby preventing unauthorized modifications by any involved stakeholders, notably regulatory agencies. This implementation was crucial for enhancing protection not only for consumers but also for producers, safeguarding their reputation in case of any downstream food-related issues and ultimately bolstering the security of the food supply chain.

As part of the ongoing work by the Royal Scientific Society (RSS) of Jordan with the Royal Academy of Engineering (RAE), UK, a novel framework has been introduced to improve RECP management and data collection at the food industry level. The framework links the IoT systems and blockchain together and was installed at the chicken breast production line in one food industry in Jordan. The main goal was to monitor product quality, environmental conditions, production process, power consumption, waste generation, and any up-normal events. The structure of the digital framework includes, (1) System typology where communication channels and type of data are programmed; (2) Data sources by which multiple data sources are identified, i.e., weighing sensor, metal sensor, temperature sensor for output meat, power consumption sensors, and shock freeze temperature sensor; and (3) IoT gateways are where the data are collected to IoT gateways and then connected to the blockchain (Fig. 3).

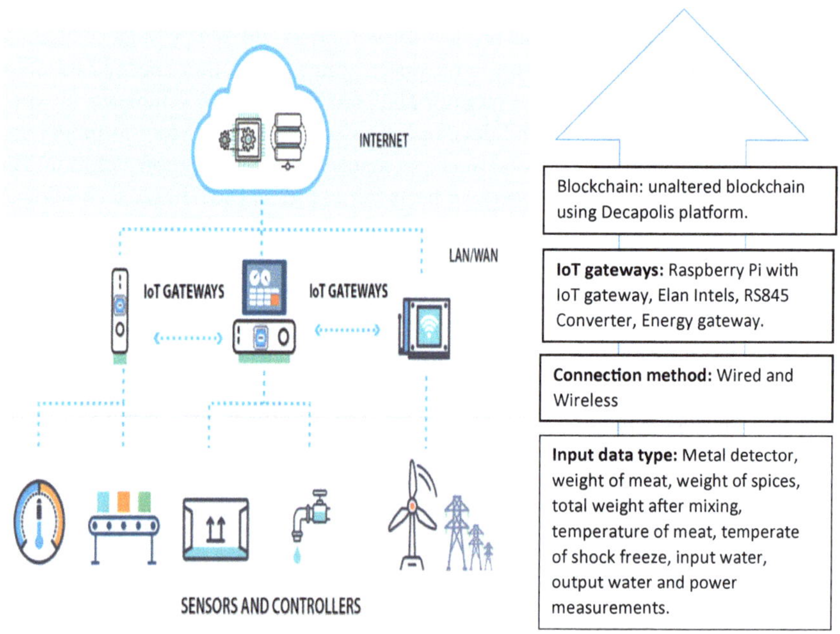

Fig. 3 The framework of IoT and blockchain that was applied at chicken breast production line at one food industry in Jordan

The data streams of energy, raw materials, water consumption, and factory emission data are then stored in an unalterable way on the Decapolis Blockchain platform. Decapolis is a Jordanian technology enterprise that provides traceability solutions for food safety using blockchain technology (https://www.decapolis.io/). This will allow the company to assess the costs of production and provide a basis for making production processes more efficient and more accurately. Figure 4 shows an example of how data is being displayed via web and mobile phone on a real-time basis and provides actual data on natural resources consumption based on the RECP assessment.

This allows the company to have an RECP database that stores transaction-based records, continually recording all completed transactions in a block of data. More importantly, this process is important to facilitate the transition to a circular economy to ensure regulatory compliance in various sectors for better processing across focal points and other actors in supply chains.

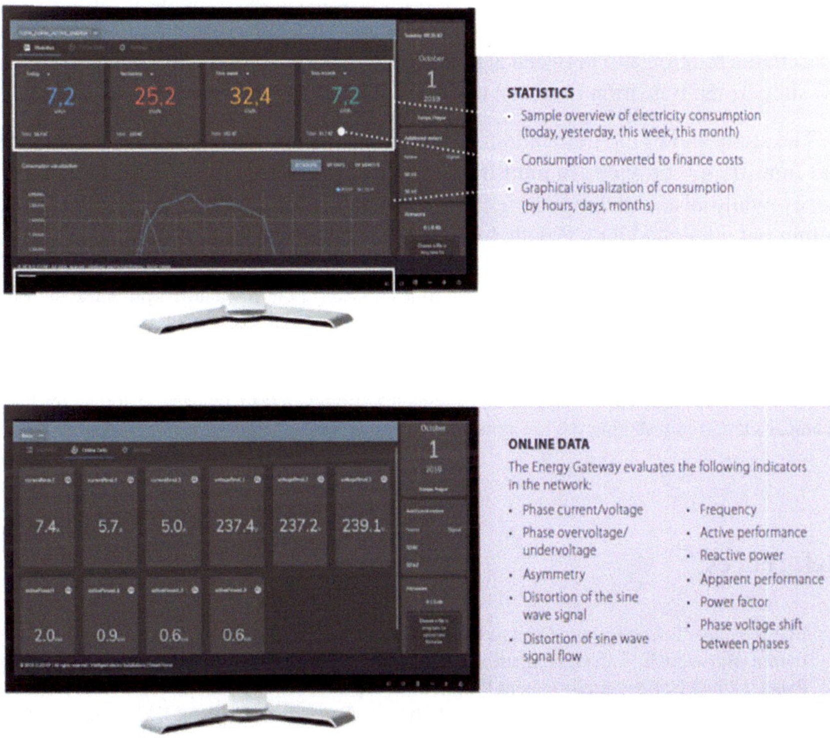

Fig. 4 Data displayed on real-time bases and stored at a blockchain software

4 Conclusions

The following conclusions can be drawn from this paper:

1. There are various approaches to implementing RECP, one out of which is TEST, which was demonstrated in 12 industries in Jordan and was behind in saving 22,181 MWh/year; 63,844 m^3/year of water and 404 tons/year of raw material.
2. The RECP measures ranged from simple measures (quick wins) to high investment measures. Yet, most of the measures were economically feasible with a pay-back period falling between 2 months and 5 years.
3. Integration of Industry 4.0 tools into RECP has improved the ability of RECP teams in industries to take timely decisions in terms of natural resource consumption. More importantly, it improved transparency, and greatly reduced inefficient manual performance evaluation and costs.
4. Blockchain will help to track data and more accurately monitor carbon footprint and ensure integrity. Linking blockchain with RECP would also establish a reliable relationship between actors without a mediator, which will significantly support the transition toward a circular economy.

The study shows the tremendous potential the integration of the RECP approach and Industry 4.0 provides to industries in helping them to improve operational efficiency while also reducing the environmental impact. We therefore recommend rolling out integrated RECP with 4.0 to other industries across all sectors.

Acknowledgements The authors would like to acknowledge the SwitchMed program, the EU-funded project, through which 12 RECP/TEST assessments were conducted at 12 industries in Jordan, in close cooperation with UNIDO. The part of integrating Industry 4.0 into RECP was funded by the Royal Academy of Engineering in the UK, Newton-Khalidi Fund, TSP program. Special thanks to Engs. Kamel and Husam Barghouthi from Gheras Company in Jordan for their technical support in installing the IoT sensors.

References

1. IRP, Resource Efficiency and Climate Change: Material Efficiency Strategies for a Low-Carbon Future. Harwich, E., Lifset, R., Pauliuk, S., Heeren, N. A report of the International Resource Panel. United Nations Environment Programme, Nairobi, Kenya, 2020.
2. Oberle B, Bringezu S, Hatfeld-Dodds S, Hellweg S, Schandl H, Clement J, Cabernard L, Che N, Chen D, Droz-Georget H, Ekins P, FischerKowalski M, Flörke M, Frank S, Froemelt A, Geschke A, Haupt M, Havlik P, Hüfner R, Lenzen M, Lieber M, Liu B, Lu Y, Lutter S, Mehr J, Miatto A, Newth D, Oberschelp C, Obersteiner M, Pfster S, Piccoli E, Schaldach R, Schüngel J, Sonderegger T, Sudheshwar A, Tanikawa H, van der Voet E, Walker C, West J, Wang Z, Zhu BA (2019) IRP, Global resources outlook: natural resources for the future we want. Report of the International Resource Panel. United Nations Environment Programme, Nairobi, Kenya
3. UNEP and IRP (2019) Natural resource use in the group of 20: status, trends and solutions
4. Jackson T (2009) Prosperity without growth-economies for a finite planet. Earthscan from Routledge, London

5. Ramos S, Etxebarria S, Cidad M, Gutierrez M, Martin DS, Iñarra B, Olabarrieta I, Melado-Herreros Á, Zufia J (2020) Cleaner production strategies for the food industry. In: Galanakis C (ed) The interaction of food industry and environment. Academic Press, pp 1–34
6. IRENA and FAO (2021) Renewable energy for agri-food systems—towards the sustainable development goals and the Paris agreement. Abu Dhabi and Rome. https://doi.org/10.4060/cb7433en
7. Mekonnen MM, Hoekstra AY (2011) National water footprint accounts: the green, blue and grey water footprint of production and consumption. Value of Water Research Report Series No. 50. UNESCO-IHE, Delft, the Netherlands
8. Henningsson S, Hyde K, Smith A, Campbell M (2004) The value of resource efficiency in the food industry: a waste minimization project in East Anglia, UK. J Clean Prod 12(5):505–512
9. UNIDO (2019) Enterprise-level indicators for resource productivity and pollution intensity a primer for small and medium-sized enterprises. https://www.unido.org/sites/default/files/2010-12/SME_Indicator_Primer_0.pdf. Accessed 12 Dec 2022
10. Fresner J, Krenn C (2021) STENUM Unternehmensberatung und Forschungsgesellschaft für Umweltfragen mbH. RECP Navigator Instruments for supporting resource efficiency and cleaner production in SMEs. https://doi.org/10.13140/RG.2.2.26088.80644
11. MedWave (2022) SwitchMed newspaper-switching to circular economy in the Mediterranean. https://switchmed.eu/wp-content/uploads/2022/12/Connect22_Newspaper_English.pdf. Accessed 24 Dec 2021
12. Med Test: Transfer of Environmental Sound Technology in the South Mediterranean Region (2012). UNIDO. https://www.unido.org/sites/default/files/2012-04/MEDTEST_%20Brochure_%20English_0.PDF. Accessed 24 Dec 2022
13. UNIDO TEST Tool Kit. Test toolkit | Unido (test-toolkit.eu). Accessed 18 Dec 2022
14. Switchmed Magazine (2018 winter). https://www.unido.org/sites/default/files/files/2020-01/SwitchMed%20Magazine%20-%20Jordan.pdf. Accessed 24 Dec 2022
15. Hasnan NZN, Yusoff Y (2018) Short review: application areas of industry 4.0 technologies in food processing sector. IEEE 16th student conference on research and development (SCOReD), Bangi, Malaysia, 26–28 Nov 2018
16. Sandeep J, Garcia-Garciac G, Rahimifarda S (2021) Optimization of the resource efficiency of food manufacturing via the internet of things. Comput Ind 127
17. Kumar V (2020) Adjusting to the new normal: challenges of the food sector in the wake of COVID-19. J Supply Chain Manag, Logist Procure 3(2):163–180
18. Khan PW, Byun Y-C, Park N (2020) IoT-blockchain enabled optimized provenance system for food industry 4.0 using advanced deep learning. Sensors (MDPI Publications) 20:2990
19. Kayikci Y, Subramanian N, Dora M, Bhatia MS (2022) Food supply chain in the era of Industry 4.0: blockchain technology implementation opportunities and impediments from the perspective of people, process, performance, and technology. Prod Plan Control 33(2–3):301–321
20. Casino F, Kanakaris V, Dasaklis TK, Moschuris S, Stachtiaris S, Pagoni M, Rachaniotis NP (2021) Blockchain-based food supply chain traceability: a case study in the dairy sector. Int J Prod Res 59(19):5758–5770
21. Kasten J (2019) Blockchain application: the dairy supply chain. J Supply Chain Manag Syst 8(1):45–54

Applications of Machine Learning in Supply Chain Management—A Review

P. Thejasree, N. Manikandan, K E K Vimal, K. Sivakumar, and P. C. Krishnamachary

Abstract Machine learning (ML) has emerged as a powerful tool in supply chain management (SCM), enabling organizations to accomplish valuable insights from numerous data and attain informed decisions. This paper presents an inclusive review of the recent advancements and applications of ML in SCM. The objective is to provide a holistic understanding of how ML techniques are being utilized to enhance various aspects of supply chain operations. The review begins by outlining the fundamental concepts of ML and its relevance to SCM. It then discusses the key challenges faced by supply chain professionals and how ML can address these challenges. The paper presents an overview of different ML techniques, including regression analysis, clustering, classification, time series analysis, neural networks, genetic algorithms, reinforcement learning, and ensemble methods, highlighting their specific applications in SCM. Furthermore, the review discusses the recent research trends and developments in the field, focusing on demand forecasting, inventory optimization, supplier selection and risk assessment, logistics optimization, supply chain risk management, and sustainability initiatives. The paper also explores the integration of ML with emergent technologies such as blockchain, IoT, and edge computing in the context of SCM. The findings of the review indicate that ML has demonstrated significant potential in improving decision-making, optimizing operations, enhancing supply chain resilience, and addressing sustainability challenges.

P. Thejasree · N. Manikandan (✉)
Department of Mechanical Engineering, School of Engineering and Technology, Mohan Babu University (MBU), Tirupati, Andhra Pradesh, India
e-mail: thuraiyurmani@gmail.com

K. E. K. Vimal
Department of Production Engineering, National Institute of Technology, Tiruchirappalli, Tamil Nadu, India

K. Sivakumar
Centre for Logistics and Supply Chain Management, Loyola Institute of Business Administration, Loyola College Campus, Chennai, TN, India

P. C. Krishnamachary
J.B. Institute of Engineering and Technology, Hyderabad, Telangana, India

K E K Vimal et al. (eds.), *Industry 4.0 Technologies: Sustainable Manufacturing Supply Chains*, Environmental Footprints and Eco-design of Products and Processes,
https://doi.org/10.1007/978-981-99-4819-2_6

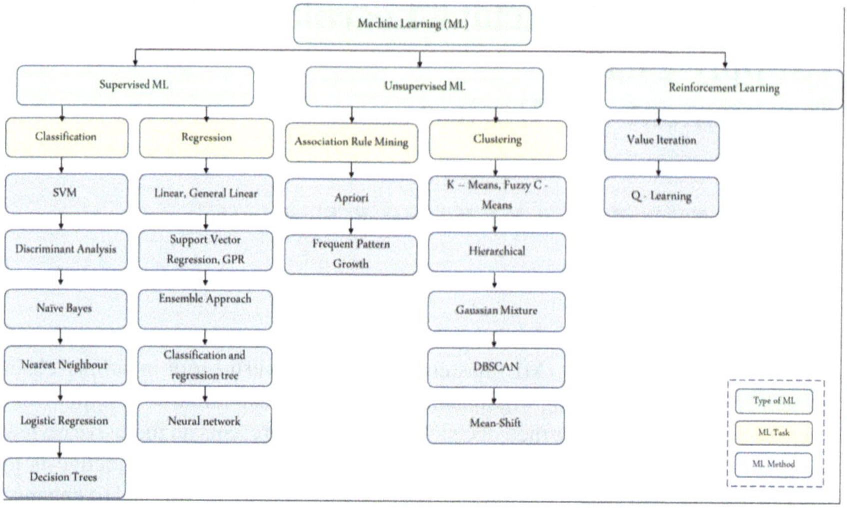

Fig. 1 Overview of machine learning methods

1 Introduction

1.1 Machine Learning (ML) Techniques

Machine learning techniques have gained significant attention and popularity in various fields because of their capability to analyze and interpret large volumes of intricate data [1, 2]. In recent years, these techniques have found widespread applications in supply chain management, revolutionizing traditional approaches to decision-making and optimization. Machine learning refers to the usage of algorithms which may absorb data and offer forecasts or decisions without being clearly programmed [3–5]. By leveraging statistical models and computational power, machine learning algorithms can recognize patterns, extract insights, and make accurate forecasts based on past and real-time data. The overview of machine learning methods has been depicted in Fig. 1 and the general system configuration of machine learning is illustrated in Fig. 2.

1.2 Supply Chain Management (SCM)

In the background of SCM, machine learning techniques offer immense potential for improving various aspects of operations, including demand forecasting, inventory management, logistics optimization, and risk assessment. These techniques enable supply chain professionals to analyze vast amounts of data from diverse sources, uncover hidden relationships, and make data-driven decisions [3].

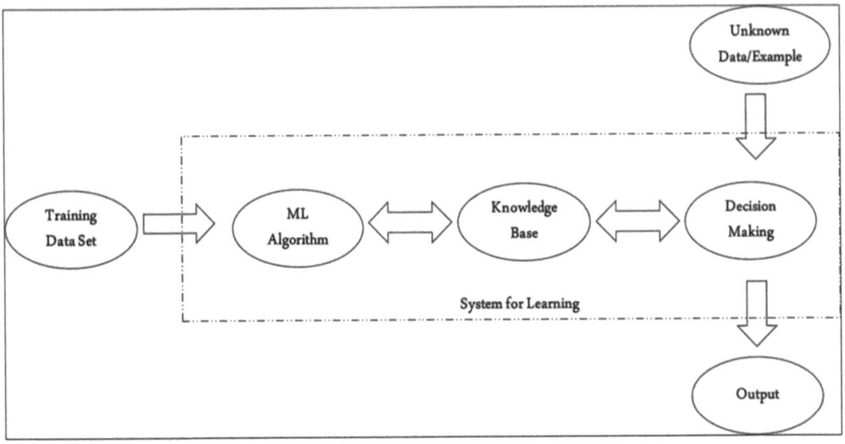

Fig. 2 Machine learning system configuration

One popular machine learning technique used in supply chain management is regression analysis, which aims to establish the correlation among dependent and independent parameters. Regression models can be adopted for sales forecasting, predicting demand patterns, and identifying influential factors that impact supply chain performance. Another widely used technique is clustering, which groups same kind of information composed based on their characteristics or attributes. Clustering algorithms have been adopted to segment customers, products, or suppliers, facilitating targeted marketing strategies, personalized recommendations, and efficient supplier management. Classification algorithms are also valuable in supply chain management, as they help categorize data into distinct classes or categories. For example, these algorithms can be used to classify products based on demand patterns, identify high-risk suppliers, or predict the likelihood of quality defects. Furthermore, time series analysis techniques, such as autoregressive integrated moving average (ARIMA) and long short-term memory (LSTM) models, are commonly used for forecasting future values based on past observations. These techniques are particularly effective in capturing temporal dependencies and seasonality patterns in supply chain data. Additionally, optimization algorithms, such as linear programming and genetic algorithms, are employed to solve complex supply chain optimization problems, such as production scheduling, inventory optimization, and route optimization [3–6].

The advent of big data and advancements in computing power has further fueled the usage of "ML" techniques in supply chain management. With the ability to process and analyze vast amounts of structured and unstructured data, machine learning algorithms can generate actionable insights, enhance decision-making processes, and enhance overall supply chain performance. Hence, machine learning techniques have emerged as powerful tools in supply chain management, enabling organizations to extract valuable insights from data, optimize operations, and gain a competitive edge. By leveraging these techniques, supply chain professionals can make more informed decisions, enhance efficiency, and adapt to dynamic market conditions. As the field of machine learning continues to advance, its applications in supply chain management are expected to evolve and drive further innovation in the industry [5–8].

2 Challenges in SCM

Supply chain professionals face various challenges in managing complex and dynamic supply chains such as demand volatility in which fluctuations in customer demand make it challenging to accurately forecast and plan inventory levels, leading to either excess inventory or stockouts [9]. Also, the lack of visibility across the supply chain hampers the ability to track inventory, monitor supplier performance, and identify bottlenecks or disruptions. Modern supply chains involve multiple tiers of suppliers, diverse product portfolios, global networks, and intricate logistics. Managing this complexity requires effective coordination and collaboration [10–12]. Balancing the level of inventories to fulfill the need of customers when minimalizing the cost of holding and stockouts is becoming one of the complex optimization problems during SCM activities. Identifying reliable suppliers, managing relationships, and assessing supplier performance is crucial for ensuring quality, timely delivery, and cost-effectiveness [13, 14].

Efficiently managing transportation, optimizing routes, and coordinating deliveries across various modes of transport can be challenging, especially with changing customer expectations and dynamic market conditions. In addition to natural disasters and political and geopolitical factors, supply chains are also vulnerable to other risks, such as supplier disruptions and quality issues. To ensure continuity, proper management and mitigation of these risks is essential. Supply chains generate massive amounts of data, including sales data, production data, logistics data, and external data. Extracting meaningful insights from this data requires robust data management and advanced analytics capabilities. Increasingly, supply chain professionals need to address sustainability goals, ethical sourcing, and environmental impacts. Balancing economic objectives with social and environmental responsibilities is a challenge. Emergent techniques such as AI, ML, blockchain, and IoT offer opportunities for innovation but require a deep understanding and strategic adoption to fully leverage their potential. Addressing these challenges requires supply chain professionals to embrace digital transformation, adopt advanced analytics and decision-support tools, foster collaboration with partners, and continuously adapt strategies to align with changing market dynamics [11–15].

3 Machine Leaning and Supply Chain Management

3.1 Need of ML in SCM

Machine learning (ML) plays a crucial role in addressing the complex challenges faced by SCM in the present days dynamic business environment [16–18]. Some key reasons highlighting the need for ML in SCM are:

Handling Big Data: SCM generates vast amounts of information from different bases such as sales transactions, production records, level of inventories, client feedback, and external factors. ML techniques are capable of processing and analyzing this big data, uncovering patterns, and extracting actionable insights for better decision-making.

Demand Forecasting: Demand forecasting is an essential part of any organization's supply chain management strategy to ensure that it is able to achieve optimal inventory levels and production planning. ML algorithms can analyze historical sales data and identify trends, seasonality, and other factors influencing demand patterns, leading to more accurate forecasts and reduced forecasting errors.

Inventory Optimization: Efficient inventory management is critical to balancing costs and customer satisfaction. ML techniques can analyze historical data, customer behavior, market trends, and other factors to optimize inventory levels, reducing stockouts and excess inventory while maximizing service levels.

Supply Chain Risk Management: Various risks can affect the operations of a supply chain, such as natural calamities, supplier disagreements, and market shifts. ML models can assess historical risk data, external data sources, and real-time information to identify potential risks, predict their impact, and enable proactive risk mitigation strategies.

Enhancing Efficiency: ML algorithms can automate and streamline routine SCM tasks, such as demand planning, order fulfillment, and logistics optimization. By reducing manual effort and human error, ML improves process efficiency, reduces lead times, and enhances overall supply chain performance.

Real-time Decision-Making: SCM requires real-time decision-making to respond to changing market conditions, customer demands, and unforeseen events. ML techniques enable organizations to process and analyze real-time data streams, providing actionable insights and supporting agile decision-making for effective supply chain management.

Customer Relationship Management: ML algorithms can analyze customer data, including preferences, buying behavior, and sentiment analysis from social media, to enhance customer relationship management. This enables targeted marketing campaigns, personalized recommendations, and improved customer satisfaction.

Supply Chain Network Optimization: ML algorithms can optimize the design and configuration of the supply chain network by considering factors such as customer demand, transportation costs, lead times, and supplier capabilities. This leads to optimized sourcing, distribution, and production strategies, improving overall supply chain efficiency.

Continuous Improvement: ML models can continuously learn and adapt to changing patterns and conditions in the supply chain. By analyzing feedback, monitoring performance, and identifying areas of improvement, ML enables continuous optimization and drives supply chain innovation.

The need for ML in SCM arises from the complexity, scale, and dynamic nature of supply chain operations. ML techniques enable organizations to leverage data-driven insights, optimize processes, mitigate risks, and enhance customer satisfaction, ultimately consequences to competent benefits in the marketplace.

3.2 Various ML Techniques Available for SCM

There are several ML techniques that can be applied to SCM to improve decision-making, optimize operations, and enhance overall performance [19–22]. Here are some commonly used ML techniques in SCM:

Regression Analysis: Regression models, such as linear regression and polynomial regression, are used for predicting and modeling relationships between dependent and independent variables. In SCM, regression analysis can be applied to forecast demand, examine the influence of factors on supply chain performance, and optimize inventory levels.

Clustering: The algorithms, such as k-means clustering and hierarchical clustering, cluster same kind of data points composed as per their characteristics or attributes. In SCM, clustering can be used for customer segmentation, product categorization, and supplier segmentation, enabling targeted strategies and personalized approaches.

Classification: It is including decision trees, random forests, and support vector machines (SVM), which are adopted to classify data into distinct classes. In SCM, classification can be applied to supplier evaluation and selection, quality control, and risk assessment, enabling effective decision-making and risk mitigation.

Time Series Analysis: Techniques such as autoregressive integrated moving average (ARIMA) and exponential smoothing, are used for forecasting future values based on past observations. In SCM, time series analysis is valuable for demand forecasting, inventory optimization, and production planning.

Neural Networks: Neural networks, including feedforward neural networks, recurrent neural networks (RNN), and long short-term memory (LSTM) networks, are used for complex pattern recognition and sequence modeling. In SCM, neural networks can be applied to demand forecasting, anomaly detection, and optimization problems, capturing nonlinear relationships and temporal dependencies in the data.

Genetic Algorithms: Genetic algorithms are optimization techniques enthused by the process of natural selection. These algorithms have been adopted to solve complex optimization problems in SCM, such as facility location, vehicle routing, and production scheduling, considering multiple constraints and objectives.

Reinforcement Learning: Reinforcement learning algorithms learn optimal decision-making policies by interacting with the environment and receiving rewards or penalties. In SCM, reinforcement learning can be applied to optimize supply chain operations, such as inventory control, order fulfillment, and routing, by dynamically adapting to changing conditions.

Support Vector Machines (SVM): This algorithm is supervised and can perform regression analysis and classification. In SCM, SVM can be applied to demand forecasting, anomaly detection, and supplier evaluation, providing accurate predictions and insights based on historical and real-time data.

Ensemble Methods: Gradient boosting and random forests are examples of ensembles that combine multiple ML models for improved accuracy and reduced overfitting. In SCM, ensemble methods can be used for demand forecasting, inventory optimization, and risk analysis, leveraging the strengths of multiple models.

These are just a few examples of ML techniques that can be applied to SCM. The selection of technique is contingent on the specific problem, available data, and desired outcomes. Often, a combination of multiple techniques and algorithms is employed to tackle the complexities and challenges of supply chain management effectively.

4 Recent Applications of ML in SCM

Recent uses of ML in SCM have shown promising results in various areas. Here are some recent applications of ML and notable research in the field of SCM [22–26]:

Demand Forecasting: ML techniques, namely neural networks and deep learning models, were applied to improve demand forecasting accuracy. Researchers have explored the use of recurrent neural networks (RNNs) and LSTM models to record temporal dependencies and seasonality patterns in demand data.

Inventory Optimization: ML algorithms have been employed to optimize inventory levels by considering demand patterns, lead times, and other factors. Reinforcement learning approaches have been used to dynamically adjust inventory policies based on real-time demand and supply information.

Supplier Selection and Risk Assessment: ML techniques have been utilized for supplier evaluation and risk assessment. Researchers have developed models that integrate various data sources, including supplier performance data, financial indicators, and external data, to identify reliable suppliers and assess their risk levels.

Logistics Optimization: ML algorithms were adopted to optimize logistics operations, such as route planning, vehicle scheduling, and load balancing. Techniques like genetic algorithms and ant colony optimization were used to find optimal solutions considering multiple constraints and objectives.

Predictive Maintenance: ML models have been utilized for predictive maintenance in SCM to improve asset reliability and decrease downtime. With the aid of analysis of sensor data, past records, and remaining information, predictive models can detect equipment failures in advance and schedule maintenance activities accordingly.

Supply Chain Risk Management: ML techniques have been used for risk identification, assessment, and mitigation in supply chains. Researchers have explored anomaly detection algorithms, such as LSTM autoencoders and support vector machines, to identify abnormal patterns in supply chain data and detect potential risks.

Blockchain and ML Integration: The integration of blockchain technology with ML has been studied for enhancing supply chain transparency, traceability, and security. Researchers have explored the use of ML algorithms for data analysis in blockchain-enabled supply chains, enabling efficient data management and decision-making.

Sustainability and Green SCM: ML techniques have been applied to support sustainability initiatives in SCM. Researchers have developed models to analyze data related to carbon emissions, energy consumption, and environmental impact, enabling organizations to optimize their supply chain operations while minimizing their environmental footprint.

Notable exploration in the field of ML applications in SCM includes studies on ensemble models for demand forecasting, hybrid optimization algorithms for supply chain network design, explainable AI for decision support in SCM, and the integration of ML with emerging technologies like Internet of Things (IoT) and edge computing for real-time analytics in SCM. Overall, the recent research in SCM focuses on leveraging ML techniques to improve decision-making, optimize operations, enhance supply chain resilience, and address sustainability challenges. These advancements contribute to the development of more intelligent, data-driven, and efficient supply chain management practices.

5 Summary

Machine learning (ML) has attained significant consideration and application in supply chain management (SCM) in recent days. It has proved to be an appreciated tool in addressing the complexities and challenges of SCM by providing data-driven insights, optimizing operations, and enhancing decision-making processes. Recent applications of ML in SCM have focused on various areas. Demand forecasting has been improved through the use of neural networks and deep learning models that can capture temporal dependencies and seasonality patterns in demand data. Inventory optimization has benefited from ML algorithms that consider demand patterns, lead times, and other factors to optimize inventory levels. ML techniques have also been applied to supplier selection and risk assessment, logistics optimization, predictive maintenance, supply chain risk management, blockchain integration, and sustainability initiatives in SCM.

Notable research in the field has explored ensemble models for demand forecasting, hybrid optimization algorithms for supply chain network design, explainable AI for decision support, and the integration of ML with emerging technologies like IoT and edge computing. These research efforts aim to improve decision-making, optimize operations, enhance supply chain resilience, and address sustainability challenges in SCM. Overall, the recent advancements in ML applications in SCM hold great potential for transforming supply chain operations, improving efficiency, reducing costs, mitigating risks, and driving sustainable practices. The integration of ML techniques with SCM practices is expected to continue growing, leading to further innovations and advancements in the field.

References

1. Ni D, Xiao Z, Lim MK (2020) A systematic review of the research trends of machine learning in supply chain management. Int J Mach Learn Cybern 11:1463–1482
2. Bousqaoui H, Achchab S, Tikito K (2017, October) Machine learning applications in supply chains: an emphasis on neural network applications. In: 2017 3rd international conference of cloud computing technologies and applications (CloudTech). IEEE, pp 1–7
3. Seif G (2018) The 5 clustering algorithms data scientists need to know. Towards Data Science
4. Du CJ, Sun DW (2006) Learning techniques used in computer vision for food quality evaluation: a review. J Food Eng 72(1):39–55
5. Wenzel H, Smit D, Sardesai S (2019) A literature review on machine learning in supply chain management. In: Artificial intelligence and digital transformation in supply chain management: innovative approaches for supply chains. Proceedings of the Hamburg international conference of logistics (HICL), vol 27. Epubli GmbH, Berlin, pp 413–441
6. Akbari M, Do TNA (2021) A systematic review of machine learning in logistics and supply chain management: current trends and future directions. Benchmarking: Int J 28(10):2977–3005
7. Hu H, Xu J, Liu M, Lim MK (2023) Vaccine supply chain management: an intelligent system utilizing blockchain, IoT and machine learning. J Bus Res 156:113480
8. Lin H, Lin J, Wang F (2022) An innovative machine learning model for supply chain management. J Innov Knowl 7(4):100276
9. Tirkolaee EB, Sadeghi S, Mooseloo FM, Vandchali HR, Aeini S (2021) Application of machine learning in supply chain management: a comprehensive overview of the main areas. Math Probl Eng 2021:1–14
10. Ghazal TM, Alzoubi HM (2022) Fusion-based supply chain collaboration using machine learning techniques. Intell Autom Soft Comput 31(3):1671–1687
11. Kohli S, Godwin GT, Urolagin S (2021) Sales prediction using linear and KNN regression. In: Advances in machine learning and computational intelligence: proceedings of ICMLCI 2019. Springer Singapore, pp 321–329

12. Shilong Z (2021, January) Machine learning model for sales forecasting by using XGBoost. In: 2021 IEEE international conference on consumer electronics and computer engineering (ICCECE). IEEE, pp 480–483
13. Park KJ (2021) Determining the tiers of a supply chain using machine learning algorithms. Symmetry 13(10):1934
14. Islam S, Amin SH (2020) Prediction of probable backorder scenarios in the supply chain using distributed random forest and gradient boosting machine learning techniques. J Big Data 7:1–22
15. Vairagade N, Logofatu D, Leon F, Muharemi F (2019) Demand forecasting using random forest and artificial neural network for supply chain management. In: Computational collective intelligence: 11th international conference, ICCCI 2019, Hendaye, France, September 4–6, 2019, Proceedings, Part I, vol 11. Springer International Publishing, pp 328–339
16. Ali MR, Nipu SMA, Khan SA (2023) A decision support system for classifying supplier selection criteria using machine learning and random forest approach. Decis. Anal. J 100238
17. Raza SA, Govindaluri SM, Bhutta MK (2023) Research themes in machine learning applications in supply chain management using bibliometric analysis tools. Benchmarking: Int J 30(3):834–867
18. Ding S, Cui T, Wu X, Du M (2022) Supply chain management based on volatility clustering: the effect of CBDC volatility. Res Int Bus Financ 62:101690
19. De Lucia C, Pazienza P, Bartlett M (2020) Does good ESG lead to better financial performances by firms? Machine learning and logistic regression models of public enterprises in Europe. Sustainability 12(13):5317
20. Li L (2022) Predicting the investment risk in supply chain management using BPNN and machine learning. Wirel Commun Mob Comput 2022
21. Sinha GK (2022) Relationship between sustainable logistics practices and the organization's performance in automobile industry—an empirical study with logistic regression machine learning. Int J Mech Eng 7(1)
22. Nguyen HD, Tran KP, Thomassey S, Hamad M (2021) Forecasting and anomaly detection approaches using LSTM and LSTM Autoencoder techniques with the applications in supply chain management. Int J Inf Manag 57:102282
23. Weng T, Liu W, Xiao J (2020) Supply chain sales forecasting based on lightGBM and LSTM combination model. Ind Manag Data Syst 120(2):265–279
24. Bousqaoui H, Achchab S, Tikito K (2019) Machine learning applications in supply chains: Long short-term memory for demand forecasting. In: Cloud computing and big data: technologies, applications and security, vol 3. Springer International Publishing, pp 301–317
25. Yani LPE, Priyatna IMA, Aamer AM (2019) Exploring machine learning applications in supply chain management. In: 9th international conference on operations and supply chain management, pp 161–169
26. Carbonneau R, Vahidov R, Laframboise K (2007) Machine learning-Based demand forecasting in supply chains. Int J Intell Inf Technol (IJIIT) 3(4):40–57

The Benefits of Using Industry 4.0 in the Manufacturing Sector

Asad Ullah, Shahid Imran, and Deena Roy

Abstract In this era of digitization, business organizations are trying to use and implement different types and levels of automation techniques for improving their efficiency and effectiveness. In this research, the researcher aims to highlight the benefits of using Fourth Industrial Revolution framework of technologies, specifically the Internet of Things (IoT) and their enabling components, in the manufacturing sector. Industry 4.0 is a combination of IoT based information sharing systems and physical computerization systems (jointly constituting the cyber-physical systems) that can help in collecting and reviewing real-time information related to the end-to-end supply chain of an organization between the starting supply point upstream and the end customers downstream. The study employs mixed methods of research for collecting, processing, and analyzing data collected from 60 employees and technicians of sampled manufacturing companies. The findings of the study suggest that employing Industry 4.0 framework may lead to an increase in productivity, improve the supply chain performance, and increase the speed of sharing information. The study recommends implementing Fourth Industrial Revolution technologies for manufacturing companies to improve and develop the total value chain, production, and cost effectiveness.

Keywords Internet of things (IoT) · Manufacturing · Technology · Fourth industrial revolution · Benefits · Supply chain management

1 Introduction

In this digital age, business organizations are adopting different types and levels of automation technologies for improving their operations. During 2011 Industry 4.0 is a framework of technologies that was conceptualized during a project of high-tech strategies by German scientists aimed at expanding the connectivity domains beyond

A. Ullah (✉) · S. Imran · D. Roy
Department of Management Studies, Middle East College, Muscat, Oman
e-mail: asadullah@mec.edu.om

K E K Vimal et al. (eds.), *Industry 4.0 Technologies: Sustainable Manufacturing Supply Chains*, Environmental Footprints and Eco-design of Products and Processes,
https://doi.org/10.1007/978-981-99-4819-2_7

the silos of proprietary communication protocols used in Supervisory Controls and Data Acquisition Systems (SCADA) and Programmable Logic Controllers (PLCs) [1]. These technologies are employed for the computerization of manufacturing operations. Industry 4.0 is a mix of IoT based open protocol (IPv6) information sharing systems and physical computerization systems that can help in collecting and reviewing real-time information related to the whole supply chain of an organization, starting from upstream suppliers to the downstream end users [2].

The data can be later used by the companies for developing effective manufacturing strategies and supply chain management (SCM) policies. The data can also be used for innovations in business processes to increase operational efficiency. It has been revealed in studies (such as [3–6]) that Fourth Industrial Revolution technologies can help organizations to enhance their productivity. For this reason, many manufacturing organizations have adopted or are considering the adoption of Industry 4.0 for managing their productions and overall business operations driving on automation [4].

According to [7], information sharing is very important in a business for managing integrity between operational units. Industry 4.0 enables the implementation of digital information systems that provides accurate information regarding the current progress of operational activities so that the business managers or the users of Industry 4.0 can take necessary initiatives for enhancing the operational efficiency of the organization [8]. In this case, if a company uses Industry 4.0, it can manage its stocks effectively by developing effective strategies for production [9]. In this process, the risks of running out of stocks can be minimized and therefore, it becomes easier for the companies to fulfill the orders of the customers. Given similar use cases, the popularity of Industry 4.0 has been increased considerably among the multinational companies as the framework helps companies to manage global supply of their products efficiently by enhancing the production of high demand and customer-focused items.

2 Aim, Objectives, and Research Questions

Aim

The basic purpose of this research is to analyze and highlight the benefits of Fourth Industrial Revolution technologies also known as Industry 4.0 in the organizations having production activities.

Objectives

(a) Understanding the concept of Fourth Industrial Revolution technologies and analyzing its benefits in manufacturing organizations.
(b) Studying the importance of implementing innovative technologies for information sharing in the manufacturing sector.

Research questions

(a) What is the concept of Fourth Industrial Revolution technologies and what are its benefits in the manufacturing sector?
(b) How is implementing innovative technologies for information sharing in the manufacturing sector important?

3 Rationale

With the purpose of increasing their productivity, many businesses are using or considering using Industry 4.0 [4]. Large manufacturing businesses, with the help of this technology, can monitor production and related organizational activities and take appropriate measures to manage logistics and supply chains [10]. Due to the difficulty of properly and timely monitoring all corporate processes without the use of real-time remote monitoring and control, the Covid-19 epidemic has prompted the necessity for digitizing and digitalizing all business operations.

As a result of it, during the pandemic situation, it had become very important for the organization to implement IoT based information system or factory monitoring system in order to develop effective strategies for enhancing operational efficiency [11]. Industry 4.0 is an IoT based automation system that computerizes all production activities and provides first-hand information regarding each stage of logistics [12]. For this reason, it has become important to analyze the impact of Industry 4.0 on the overall performance of business organizations so that companies can analyze how its technologies are contributing to the enhancement of productivity and what challenges an organization might face while using them. Through this analysis, companies may prepare for implementing Industry 4.0 and can adopt preventive measures to mitigate the possible risks associated with the implementation of Industry 4.0.

4 Significance of the Research

This study aims to help analyze the impact of the introduction of Industry 4.0 in manufacturing. In this context, various research papers related to Industry 4.0 were reviewed to gain in-depth knowledge of the technology's use in various fields. Further, the results of a survey among 60 employees and technicians of sampled manufacturing companies are presented. The pandemic of Covid-19 and the resulting lockdowns have created the need for digitization and digitalization of industrial operations [13]. The digitized and digitalized computerization of production systems has become very important. This study will serve as a reference for studies analyzing how Industry 4.0 can help companies improve their productivity.

5 Literature Review

Reviewing relevant research literature helps researchers gain in-depth knowledge of their field. For this reason, it is very important to review the relevant research literature before starting data collection and subsequent research activities. The purpose of this research work is to analyze the importance of Fourth revolution technologies in the manufacturing sector. Therefore, to understand the existing research findings on the impact of Fourth revolution technologies in manufacturing, we selected relevant research literature, such as books, book chapters, and prominent journal articles related to Industry 4.0. Additionally, the reviewed literature helped to understand the benefits for manufacturing companies looking to implement Fourth revolution technologies. The most important findings from the reviewed literature are briefly summarized in this section.

5.1 *Industry 4.0 in Manufacturing*

4IR technologies are driving the recent trends in production and manufacturing [14]. The technology framework consists of IoT-enabled sensors, cyber-physical systems, big data analytics, cloud computing, and machine learning-based artificial intelligence to effectively share information and improve manufacturing operations within various manufacturing company departments. It will be realized as an intelligent streamlined factory (Fig. 1).

The foundation of 4IR technologies infrastructure consists of cyber-physical systems and Industrial Internet of Things (IIoT) comprising mainly of smart machines, smart equipment, and smart robotics [16]. The main usage of cyber-physical systems is as sensors of processing events for the cloud-hosted control systems for tracking different production and logistics engineering activities in the whole value chain of manufacturing organizations [16, 17]. On the other hand, inbuilt IIoT based technologies are utilized for information sensing, consolidation, sharing, and analytics enabling effective management of available resources leading to new ways of production. Thus, it helps manufacturing companies to create more value for their customers, investors, and other stakeholders.

4IR technologies offer real-time data on the entire manufacturing process, which may be utilized to measure manufacturing performance metrics in real-time [18]. The users of analytics information can view the organization's value chain, assuring superior values for the chain's operational activities. Real-time monitoring is possible for a variety of events, including the materials used in production, their delivery through various stages, their origin, other production-related activities, and more.

It is easier for production firms to have strategies for controlling the supply chain and increasing the firm's production rate when the entire value chain and production operations are tracked and monitored in real-time [19]. The production operation can be properly managed by effectively controlling the resources and the distribution of

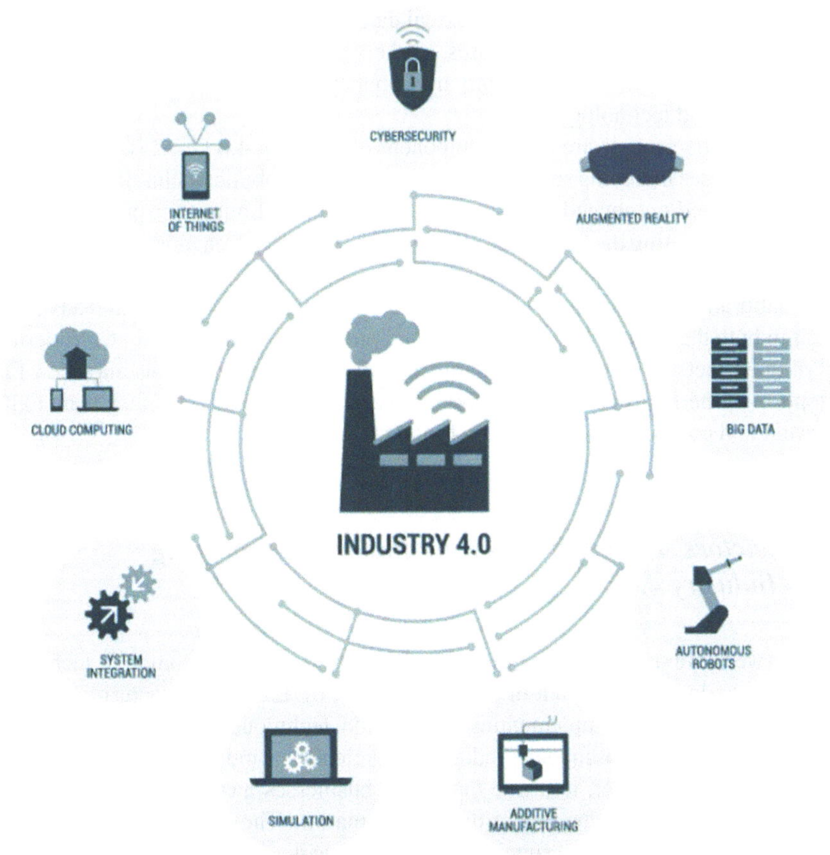

Fig. 1 Various facets of 4IR technologies. *Source* Strandhagen et al. [15]

items. Because of this, research studies (like [3, 4, 6]) have shown that the effective application of the Indindustry0 framework can aid in increasing the productivity of industrial firms. Moreover, by effectively and efficiently responding to demand fluctuations, a business can use increased production to cater to the orders of consumers in various market segments.

Similarly, by carefully planning and implementing the Industry 4.0 framework, the hazards of excess production and the resource loss they bring about can also be avoided [20]. The ability to manage a complete digital supply chain can be improved by implementing the Industry 4.0 architecture. One of the user interface elements of Industry 4.0, augmented reality, acts as the interface for end-to-end management [21]. According to [22], big businesses use augmented reality tools like immersive helmets and glasses to build smart factories because they see augmented reality as a tool for improving staff communication as well as production monitoring and control. This is based on the idea that increasing worker productivity may be aided by the visualization of operational and contextual data. According to information obtained from various manufacturing businesses, workers who are responsible for maintaining and repairing machinery, equipment, and robots employ these kinds of augmented reality tools and technology [23].

Collaborative robots are a key component of Industry 4.0 [24, 25]. In manufacturing processes involving several robots, collaborative robotics technology can carry out complicated automated tasks with little likelihood of human error while monitoring and regulating the processes. HR can now concentrate on more productive and innovative business sectors as complex industrial processes are automated. The use of collaborative robotics in additive manufacturing represents a substantial advancement in both the engineering of materials and production methods [26]. Considering all of these facts, it is apparent that large manufacturing organizations may view the adoption of the Industry 4.0 framework as necessary in the future if they want to be relevant and competitive in their target markets.

5.2 *Factors Behind the Growing Importance of Using Industry 4.0*

In the twenty-first century, the digital revolution has made automotive technologies more and more important in production [3, 4, 6]. Large manufacturing companies are interested in implementing automation techniques to improve information exchange and consequent production efficiency in industrial communication networks [27]. However, it is true that small businesses have not shown adequate interest in embracing technologies digital automation. The impact of the Covid-19 pandemic on production and supply chains has made it urgent to digitize all business processes [15]. Manufacturing activities of various organizations declined as most organizations were forced to manage manufacturing and supply chain operations with minimal numbers of workers at their factories during lockdowns designed to contain the spread of the coronavirus is severely affected. The output and efficiency of many small organizations declined during the lockdown period [28]. As a result, it has become imperative that manufacturing departments use automation technology to maintain targeted yields and efficiencies and control the response to demand. This allows companies to fulfill target customer orders in their market segments.

It became essential for manufacturing organizations to build digital supply chains for taking care of the flow of essential materials and products in the supply chain, during the Covid-19 pandemic as there was a very significant boom in E-Commerce. For managing the quality of production and maintaining efficiency with regulated responsiveness to demand, supply chain can be implemented and monitored digitally with the use of fourth industrial revolution technologies [7, 28]. This has been done by many manufacturing firms to manage their digital supply chains and e-commerce [29].

Apart from this the implementation of the 4th industrial revolution technology also assists in tracking and managing the distribution of goods in different niche markets by developing customer—oriented data analytics and driving analytics—driven production capabilities [18]. The automotive technologies and usage of robotics help in performing repetitive tasks effectively with guaranteed quality and performance outcomes in the manufacturing organizations based on several analytical capabilities, such as anomaly detection, smart maintenance, predictive quality analytics, predictive drilling efficiency, and predictive manufacturing accuracy with costs [18, 24]. It reduces organizational labor costs and increases organizational productivity [29]. Machine learning and artificial intelligence play an important role in automated analysis capabilities using regression analysis methods, support vector machines, neural networks, and decision trees [18]. Automation through machine learning and artificial intelligence makes it easier for manufacturing companies to manage production rates with a minimal workforce. For this reason, the relevance of Industry 4.0 is growing in the post-Corona business world.

It is because of these factors the importance of Industry 4.0 is growing after the post-Covid business world. Another reason cited according to [30] is the high rate of population growth leading to the implementation of Industry 4.0. This is so because in many developing countries, due to the high growth of population, the demand for various types of products has increased, specifically for food items and beverages, clothing, and health care products. Due to the implementation of Industry 4.0 technologies, manufacturing firms are able to produce massive quantities of goods, which helps them in catering to the choices of their customers with guaranteed and consistent quality and performance. It also allows the manufacturing firms to fulfill the orders of target customers with flexibility, accurately and timely. Some of the other benefits of implementing industry 4.0 technologies cited by [31] are that it helps in reducing defects and errors in automated manufacturing operations. In addition, the sharing of effective information through industry 4.0 technologies like IoT, augmented reality, etc. helps in increasing the productivity of workers and facilitates effective strategic management for the value creation process management [4, 32]. Thus, it can be said that, considering the findings of these empirical studies, over the period the importance of Industry 4.0 has increased in the manufacturing sector due to several factors but the most important reason was the spread of Covid-19 pandemic, which forced many of the manufacturing firms to implement industry 4.0.

5.3 Benefits of Adopting Industry 4.0

5.3.1 Increase of Productivity

It has been shown in studies [5, 32–34] Industry 4.0 implementations can help improve the productivity of manufacturing organizations. The timely and accurate exchange of information using IoT, and cloud-based technologies helps companies effectively audit each stage of the supply chain. This process facilitates organizations to develop supply chain and production control strategies [6]. Your organization's repetitive tasks can be executed with guaranteed quality and performance. As a result, you can produce more goods in less time while managing output, quality, performance, and customer orientation.

5.3.2 Remaining Competitive

The introduction of Industry 4.0 will enable industries to develop customer-centric, demand-driven, and dynamic capacity without sacrificing output [35, 36]. These skills, coupled with the timely, relevant, and accurate management of production, supply chain and distribution strategies according to customer requirements, will enable companies in today's highly competitive and dynamic markets targeted by manufacturing industries helps you stay competitive.

5.3.3 Increased Knowledge Sharing and Collaborative Working

Industry 4.0 includes IoT and cloud-based information exchange technologies that support the sharing of relevant information inside and outside manufacturing companies, manage collaboration and communication accordingly [37], and Helping the flow of timely, relevant, and accurate information, contractors and suppliers improve team collaboration. A real-time flow of information about supply chain events, manufacturing events, sales, and shipments helps manufacturing and sales teams work together to develop strategies that improve product quality, customer satisfaction, and company profitability increases.

5.3.4 Cost Effectiveness

Economic efficiency is another benefit that Industry 4.0 offers. Industry 4.0 technologies can automate many production and supply chain activities. Smart technology can be used to perform repetitive tasks with guaranteed performance and quality [27]. The probability of human error remains low in this process. Additionally, organizations can also manage production targets with a limited number of workers. As a result, the company's labor costs are also reduced.

5.3.5 Flexibility and Agility

Industry 4.0 implementations also help manufacturing systems remain flexible and agile. Implementing Industry 4.0 makes it easier to scale up and down production or add new product lines, depending on demand dynamics [35]. Industry 4.0 also opens up opportunities for his one-off and multi-variety manufacturing systems for highly customized products manufactured according to the customer's personal preferences. Moreover, when carefully designed, manufacturing and supply chain systems can tolerate financial, information, technical, safety, human, legal, environmental, operational, and requirements risks [30].

5.3.6 Improved Satisfaction of the Customer

The usage of intelligent machines and automotive technology in making the products can help improve the quality of the product as it can reduce the likelihood of defects and errors in product manufacturing [38]. By ensuring the quality and performance of our production processes and products, we can manufacture large quantities of products to exact and targeted specifications in limited time frames. Improving product quality and customer-focused production helps improve customer satisfaction. Management of on-demand production volumes, machine, equipment and robot uptime, and rapid redesign and readjustment of production based on fluctuations in demand help companies manage product availability in various markets. Be helpful and make sure your customers get the products they want and when they want them.

6 Research Methodology

In this study, we have followed the research onion framework, which provides clear steps for choosing research methods and conducting research activities. There are several layers in the research onion as shown in the figure below and each layer indicates a set of procedures for conducting research. Hence, the first step of the research onion framework is, selecting an appropriate research philosophy was mandatory. In this research, we have adopted the positive research philosophy which focuses on using pre-existing secondary and primary data collection from human participants and using those data statistically to obtain accurate results. In addition to making the choice of appropriate research philosophy, selecting a suitable research approach is also very crucial. Therefore, in this study survey and mixed methods of research have been used (Fig. 2).

The overall strategy of conducting research is known as research design, which is of mainly three types. These are descriptive, exploratory, and causal research design [40]. In this study, the researchers have chosen a descriptive research design, that attempts to analyze thoroughly the various aspects of research for identifying the problems and finding possible solutions for solving those problems [41]. This study also makes use of both qualitative and quantitative data for achieving the set goals of the research work. The quantitative data collected through the questionnaire survey helped in understanding better the viewpoints of respondents regarding Industry 4.0 and helped in understanding the difficulties while implementing Industry 4.0 [42]. On the other hand, the primary and secondary sources of data which were qualitative in

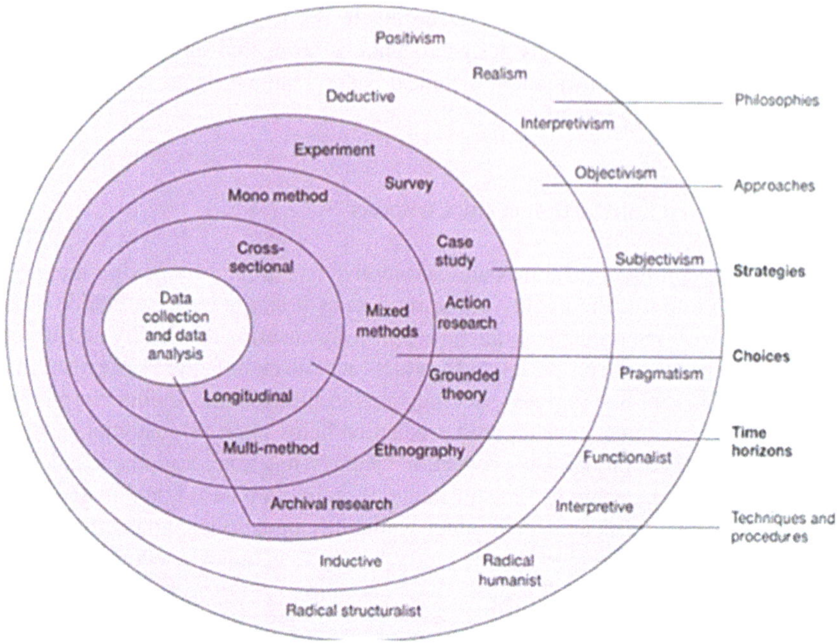

Fig. 2 Research onion. *Source* [39]

nature helped in developing detailed knowledge about the impact of 4IR technologies in the manufacturing sector [43].

6.1 Data Collection Methods

As stated above, both qualitative and quantitative data have been used in this study. The quantitative data has been collected through the questionnaire survey while the quantitative data has been collected through the secondary sources. For collecting the primary data, interviews with some managers of the manufacturing companies were also conducted. In this study, we used Google Forms as a tool to conduct an online survey of manufacturing company employees and technicians. In addition to collecting primary data, secondary qualitative data were also collected. Various scientific journals, books, articles, and websites were used as secondary sources [44]. The study also analyzed several cases of technology implementation in the manufacturing industry to gain in-depth knowledge about the impact of 4IR technologies in manufacturing. In this research, the population for conducting the survey comprised of the managers and technicians working in the manufacturing industry in Muscat. The number of respondents that were contacted to collect the responses was 70, out of which 60 responses were received. The study adopted one of the non-probability sampling techniques called purposive sampling which is also known as judgmental sampling. This type of sampling technique helps the researcher in relying on their judgments while making the choice of respondents to participate in the survey [45].

7 Data Analysis and Discussion

Choosing the right data analysis method is very important for obtaining accurate research results. In this study, primary data have been collected through online surveys which were analyzed using charts, tables, and spreadsheets. This process facilitates the analysis of levels of agreement and disagreement on various factors related to the adoption of 4IR technologies in production companies.

7.1 Profile of the Respondents

In this study, the responses from 60 respondents were considered for ascertaining the results and findings. The respondents belonged to the category of managers and technicians working in the manufacturing companies having supply in and across Muscat. Most of the respondents who participated in this survey belonged to the age group of 26–33 years having working experience of 1–3 years. Furthermore, the majority of the respondents were male. The reliability of the questionnaire was ascertained using Cronbach's Alpha, which according to [46], refers to the mean consistency of 'q' measures that includes 'q' raters, alternative forms, or questionnaire/test items.

From the various studies and reports it has been found out the acceptable range of alpha value should be from 0.70 and above, thus in this research the Cronbach credibility was found out to be alpha = 0.733 meaning that the distribution of the questionnaire was acceptable (please see the details in the Annexure).

7.2 Findings

From the analysis of the data, it has been found that majority of the respondents about (50%), were of the view that if the manufacturing companies adopt industry 4.0 technologies it will be beneficial for the manufacturing industry. Not only that specifically it will help the companies in improving the productivity. It is clearly also evident that about (50%) of the respondents agreed that IoT based information sharing technologies will help the organizations in better managing their supply chains and the exact movement of goods and materials.

On the supplier's side from the analysis of the data, it is evident that majority of the respondents strongly agreed with the view that implementation and adoption of IoT based information sharing technologies would help in improving the communication with the suppliers in terms of tracking the movements of goods and managing the supply of goods in different market segments. Yet, another benefit that was identified regarding the adoption of Industry 4.0 was that it can help the production companies in reducing the labor cost. Thus, it can be said that implementation of Industry 4.0 can help the production companies in improving their productivity, and better information sharing by providing real-time information on the movement of materials and goods, thereby making it easier for companies to effectively manage their supply chains (please refer to the Annexure for more details).

8 Conclusion

From the overall results of our research work, we can conclude that the term Industry 4.0 is becoming more and more popular in the digital age. Industry 4.0 refers to the digital transformation of production processes or the implementation of smart factories. The adoption of Industry 4.0 helps manufacturing companies to improve their production rates given the market demand for their products. Importantly, manufacturing companies must use Industry 4.0 platforms to manage the production and delivery of different products in different market segments. But mobile technology, intelligent machines based on artificial intelligence, and the Internet of Things can also contribute to the implementation of the digital supply chain. Digital supply chains help companies use smart objects to automatically perform supply chain management tasks. Additionally, the Covid-19 pandemic has accelerated the adoption of Fourth Industrial Revolution technologies.

However, the adoption of Fourth Industrial Revolution technologies can also lead to cyber security risks, as cloud computing technologies, the Internet of Things, and digital information systems are vulnerable to cyber-attacks. Therefore, companies adopting Industry 4.0 need to improve their cybersecurity infrastructure to mitigate cybersecurity risks. It also increases the cost of implementing advanced digital technology.

9 Recommendations

(a) The implementation of Fourth Industrial revolution technologies makes it easier for production companies to improve and develop value, production rate, and cost efficiency across the chain, helping manufacturing companies to fulfill customer orders in various market segments.
(b) The development of shared systems can help organizations maintain distribution processes by tracking goods and resources produced in the market.

(c) The use of new and innovative automation technologies will help organizations reduce labor costs, provide real-time information to markets and supply chains, implement digital supply chains and manage systems, and improve supply chain performance.
(d) According to recent research, Industry 4.0 may increase cybersecurity risks for manufacturing companies. Therefore, organizations interested in adopting or leveraging Industry 4.0 should hire specialists who can track activity on the company's servers and detect new cybersecurity threats. Companies can improve their corporate cybersecurity framework with the help of professional IT professionals. In some cases, an organization's administrators fail to properly handle advanced technology, and administrator error can lead to hackers gaining access to sensitive company information. For this reason, companies using advanced technology also need to train their managers to use technology properly and avoid hackers.

10 Future Research Direction

The existing literature does not identify which types of cybersecurity risks are very common for organizations deploying Industry 4.0. At the same time, the study did not find any effective solutions to help avoid cybersecurity risks arising from the use of modern digital technologies. Challenges and risks should be examined. Furthermore, effective solutions that can mitigate these risks should also be identified in future research projects.

Annexure

Test of Reliability of Questionnaire

Case processing summary			
		N	%
Cases	Valid	60	100.0
	Exclude[a]	0	0.0
	Total	60	100.0

[a]Listwise deletion based on all variable in the procedure

Reliability statistics	
Cronbach's Alpha	N of items
0.733	12

Questions

1. Do you think that Industry 4.0 is beneficial for the growth and sustainability of the manufacturing companies? (Table 1 and Fig. 3)

2. Do you believe that the implementation of automation technologies and smart machines can help enhance the productivity of the manufacturing companies? (Table 2 and Fig. 4)

3. Do you believe that the IoT based information sharing technologies can help organizations to manage their supply chain? (Table 3 and Fig. 5)

4. Do you believe that use of IoT technologies is important for information sharing regarding the movement of goods and materials? (Table 4 and Fig. 6)

5. Do you feel that the implementation of IoT based information sharing technologies can help manage communication with suppliers? (Table 5 and Fig. 7)

Table 1 Benefit of Industry 4.0

Provided options	Response received	Percentage (%)
Strongly agree	30	50
Agree	10	16.7
Neutral	10	16.7
Disagree	6	10
Strongly disagree	4	6.7

Source Created by self

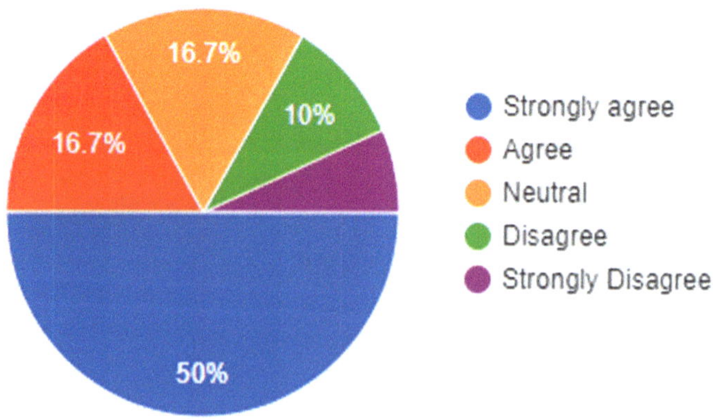

Fig. 3 Benefit of Industry 4.0

Table 2 Increase of productivity

Provided options	Response received	Percentage (%)
Strongly agree	35	58.3
Agree	10	16.7
Neutral	5	8.3
Disagree	7	11.7
Strongly disagree	3	5

Source Created by self

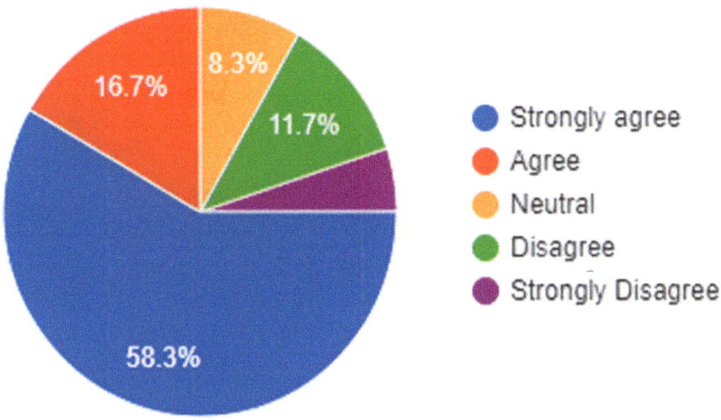

Fig. 4 Increase of productivity

Table 3 Supply chain management

Provided options	Response received	Percentage (%)
Strongly agree	30	50
Agree	15	25
Neutral	8	13.3
Disagree	3	5
Strongly disagree	4	6.7

Source Created by self

6. Do you believe that implementation of IoT based information sharing technologies are important for tracking the movement of goods and managing the supply of goods in different market segments? (Table 6 and Fig. 8).

7. Do you feel that the adoption of Industry 4.0 can help reduce the labor cost of manufacturing organizations? (Table 7 and Fig. 9)

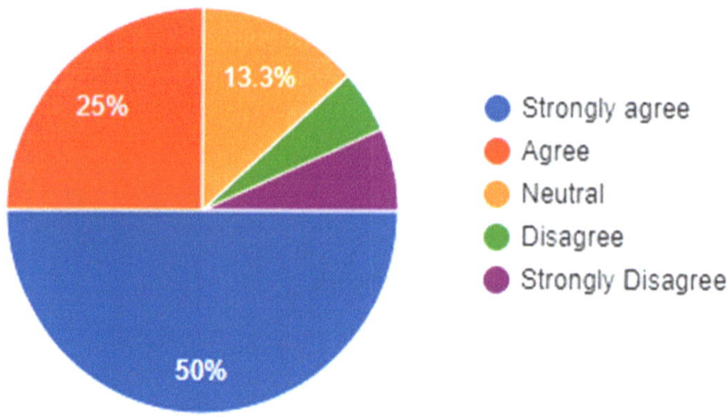

Fig. 5 Supply chain management

Table 4 Goods and materials movement

Provided options	Response received	Percentage (%)
Strongly agree	30	50
Agree	10	16.7
Neutral	10	16.7
Disagree	5	8.3
Strongly disagree	5	8.3

Source Created by self

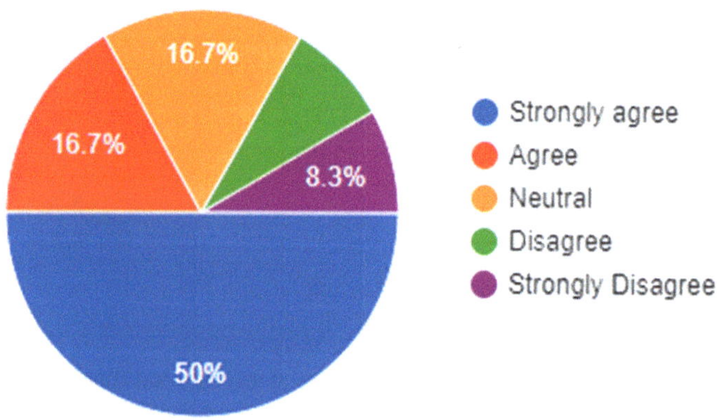

Fig. 6 Goods and materials movement

Table 5 Communication with suppliers

Provided options	Response received	Percentage (%)
Strongly agree	24	40
Agree	20	33.3
Neutral	6	10
Disagree	8	13.3
Strongly disagree	2	3.3

Source Created by self

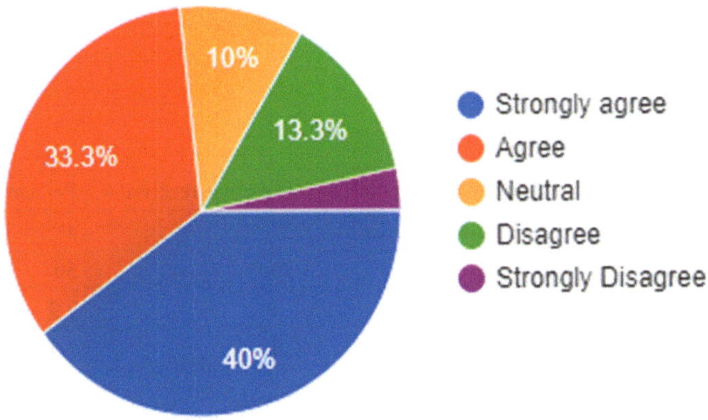

Fig. 7 Communication with suppliers

Table 6 Tracking the movement

Provided options	Response received	Percentage (%)
Strongly agree	35	58.3
Agree	15	25
Neutral	8	13.3
Disagree	1	1.7
Strongly disagree	1	1.7

Source Created by self

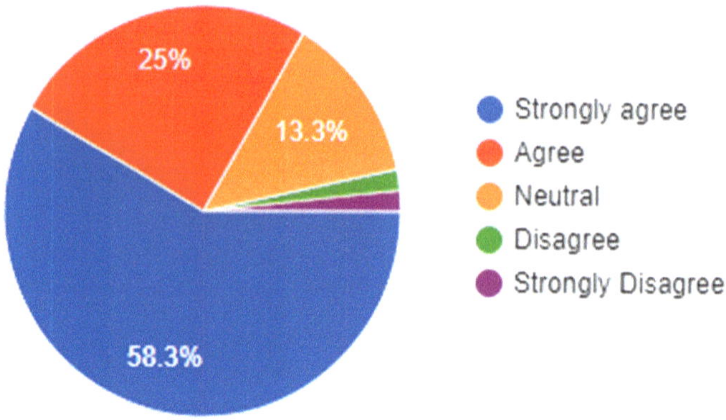

Fig. 8 Tracking the movement

Table 7 Decrease of labor cost

Provided options	Response received	Percentage (%)
Strongly agree	25	41.7
Agree	30	50
Neutral	2	3.3
Disagree	2	3.3
Strongly disagree	1	1.7

Source Created by self

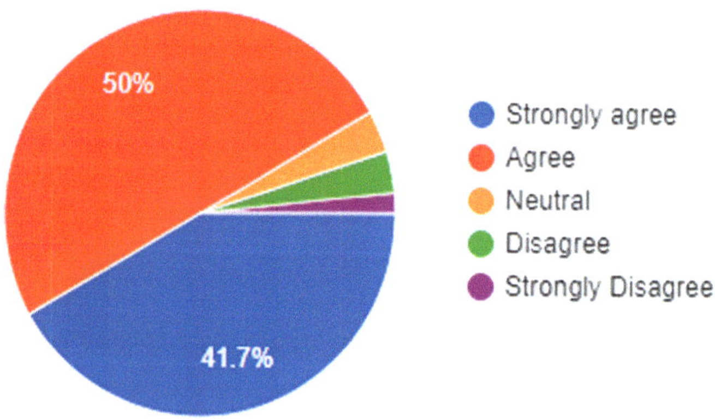

Fig. 9 Decrease of labor cost

References

1. Frank AG, Dalenogare LS, Ayala NF (2019) Industry 4.0 technologies: implementation patterns in manufacturing companies. Int J Prod Econ 210:15–26
2. Zhong RY, Xu X, Klotz E, Newman ST (2017) Intelligent manufacturing in the context of industry 4.0: a review. Engineering 3(5):616–630
3. Horvat D, Kroll H, Jager A (2019) Researching the effects of automation and digitalization on manufacturing companies' productivity in the early stage of industry 4.0. In: 25th international conference on production research manufacturing innovation: cyber physical manufacturing, 9–14 August 2019, Chicago, Illinois (USA). Procedia Manuf. 39:886–893, Elsevier
4. Ortiz JH, Marroquin WG, Cifuentes LZ (2020) In: Ortiz JH (ed) Industry 4.0: current status and future trends. IntechOpen, London, UK, pp 13–28
5. Sader S, Husti I, Daroczi M (2019) Quality management practices in the era of Industry 4.0. Zarządzanie Nr 35:117–126, Zeszyty Naukowe Politechniki Częstochowskiej
6. Schuh G, Potente T, Wesch-Potente C, Weber AR, Prote J (2014) Collaboration mechanisms to increase productivity in the context of Industrie 4.0. Procedia CIRP 19:51–56, ScienceDirect
7. Raj A, Dwivedi G, Sharma A, de Sousa Jabbour ABL, Rajak S (2020) Barriers to the adoption of industry 4.0 technologies in the manufacturing sector: an inter-country comparative perspective. Int J Prod Econ 224:107546
8. Fatorachian H, Kazemi H (2018) A critical investigation of Industry 4.0 in manufacturing: theoretical operationalisation framework. Prod Plan Control 29(8):633–644
9. Lee J, Davari H, Singh J, Pandhare V (2018) Industrial artificial intelligence for industry 4.0-based manufacturing systems. Manuf Lett 18:20–23
10. Ghobakhloo M (2018) The future of manufacturing industry: a strategic roadmap toward Industry 4.0. J Manuf Technol Manag
11. Castelo-Branco I, Cruz-Jesus F, Oliveira T (2019) Assessing industry 4.0 readiness in manufacturing: evidence for the European Union. Comput Ind 107:22–32
12. Ludbrook F, Michalikova KF, Musova Z, Suler P (2019) Business models for sustainable innovation in industry 4.0: smart manufacturing processes, digitalization of production systems, and data-driven decision making. J Self-Gov Manag Econ 7(3):21–26
13. Czifra G, Molnár Z (2020) Covid-19 and industry 4.0. Res Pap Fac Mater Sci Technol Slovak Univ Technol 28(46):36–45
14. Masood T, Sonntag P (2020) Industry 4.0: adoption challenges and benefits for SMEs. Comput Ind 121:103261
15. Strandhagen JW, Alfnes E, Strandhagen JO, Vallandingham LR (2017) The fit of industry 4.0 applications in manufacturing logistics: a multiple case study. Adv Manuf 5(4):344–358
16. Salkin C, Oner M, Ustundag A, Cevikcan E (2018) A conceptual framework for industry 4.0. In: Ustundag A, Cevikcan E (eds) Industry 4.0: managing the digital transformation. Springer International Publishing, Switzerland, pp 3–23
17. Nikolic B, Ignjatic J, Suzic N, Stevanov B, Rikalovic A (2017) Predictive manufacturing systems in industry 4.0: trends, benefits and challenges. In: Annals of DAAAM and proceedings, p 28
18. Sivri MS, Oztaysi (2018) Data analytics in manufacturing. In: Ustundag A, Cevikcan E (eds) Industry 4.0: managing the digital transformation. Springer International Publishing, Switzerland, pp 155–172
19. Ghobakhloo M, Fathi M (2019) Corporate survival in Industry 4.0 era: the enabling role of lean-digitized manufacturing. J Manuf Technol Manag
20. Butt J (2020) A strategic roadmap for the manufacturing industry to implement industry 4.0. Designs 4(2):11
21. Esengun M, Ince G (2018) The role of augmented reality in the age of industry 4.0. In: Ustundag A, Cevikcan E (eds) Industry 4.0: managing the digital transformation. Springer International Publishing, Switzerland, pp 201–215

22. Moktadir MA, Ali SM, Kusi-Sarpong S, Shaikh MAA (2018) Assessing challenges for implementing industry 4.0: implications for process safety and environmental protection. Process Saf Environ Prot 117:730–741
23. Machado CG, Winroth MP, Ribeiro da Silva EHD (2020) Sustainable manufacturing in industry 4.0: an emerging research agenda. Int J Prod Res 58(5):1462–1484
24. Bayram B, Ince G (2018) Advances in robotics in the era of industry 4.0. In: Ustundag A, Cevikcan E (eds) Industry 4.0: managing the digital transformation. Springer International Publishing, Switzerland, pp 187–200
25. Sharma M, Kamble S, Mani V, Sehrawat R, Belhadi A, Sharma V (2021) Industry 4.0 adoption for sustainability in multi-tier manufacturing supply chain in emerging economies. J Clean Prod 281:125013
26. Beyca OF, Hancerliogullari G, Yazici I (2018) The role of augmented reality in the age of industry 4.0. In: Ustundag A, Cevikcan E (eds) Industry 4.0: managing the digital transformation. Springer International Publishing, Switzerland, pp 217–234
27. Wollschlaeger M, Sauter T, Jasperneite J (2017) The future of industrial communication: automation networks in the era of the internet of things and industry 4.0. IEEE Electron Mag (IEEE) 17–27
28. Kazemi H (2018) A critical investigation of industry 4.0 in manufacturing: theoretical operationalisation framework. Prod Plan Control 29(8):633–644.
29. Zheng T, Ardolino M, Bacchetti A, Perona M, Zanardini M (2019) The impacts of industry 4.0: a descriptive survey in the Italian manufacturing sector. J Manuf Technol Manag
30. Kolla S, Minufekr M, Plapper P (2019) Deriving essential components of lean and industry 4.0 assessment model for manufacturing SMEs. Procedia CIRP 81:753–758
31. Luthra S, Mangla SK (2018) Evaluating challenges to industry 4.0 initiatives for supply chain sustainability in emerging economies. Process Saf Environ Prot 117:168–179
32. Horvath D, Szabo RZ (2019) Driving forces and barriers of Industry 4.0: do multinational and small and medium-sized companies have equal opportunities? Technol Forecast Soc Chang (Elsevier) 146:119–132
33. Mohamed M (2018) Challenges and benefits of industry 4.0: an overview. Int J Supply Oper Manag 5(3):256–265
34. Ortiz JH, Marroquin WG, Cifuentes LZ (2020) In: Ortiz JH (ed) Industry 4.0: current status and future trends. IntechOpen, London, UK, pp 13–28
35. Bibby L, Dehe B (2018) Defining and assessing industry 4.0 maturity levels—case of the defence sector. Prod Plan Control 29(12):1030–1043
36. Choe MJ, Noh GY (2018) Combined model of technology acceptance and innovation diffusion theory for adoption of smartwatch. Int J Contents 14(3):32–38
37. Schumacher A, Schumacher C, Sihn W (2020) Industry 4.0 operationalization based on an integrated framework of industrial digitalization and automation. In: Durakbasa NM, Gencyilmaz MG (eds) ICPR1 2019, LNME. Springer Nature Switzerland AG, pp 301–310
38. Tupa J, Simota J, Steiner F (2017) Aspects of risk management implementation for industry 4.0. Procedia Manuf (Elsevier) 11:1223–1230
39. Prinsloo J, Sinha S, von Solms B (2019) A review of industry 4.0 manufacturing process security risks. Appl Sci 9(23):5105
40. Melnikovas A (2018) Towards an explicit research methodology: adapting research onion model for futures studies. J Futur Stud 23(2):29–44
41. Rahi S (2017) Research design and methods: a systematic review of research paradigms, sampling issues and instruments development. Int J Econ Manag Sci 6(2):1–5
42. Grube D, Malik AA, Bilberg A (2017) Generic challenges and automation solutions in manufacturing SMES. In: Annals of DAAAM and proceedings, p 28
43. Bogna F, Raineri A, Dell G (2020) Critical realism and constructivism: merging research paradigms for a deeper qualitative study. Qual Res Organ Manag: Int J
44. Apuke OD (2017) Quantitative research methods: a synopsis approach. Kuwait Chapter Arab J Bus Manag Rev 33(5471):1–8

45. Allan G (2020) Qualitative research. In: Handbook for research students in the social sciences. Routledge, pp 177–189
46. Bonett D, Wright T (2014) Cronbach's alpha reliability: interval estimation, hypothesis testing, and sample size planning. J Organ Behav 36(1):3–15

Application of Digital Twin for Efficient Supply Chain: Analysis of Opportunities and Challenges

Laxmi Pandit Vishwakarma and Rajesh Kr Singh

Abstract Digital Twins are the simulated models of a supply chain, which use real-time data to forecast the supply chain dynamics. The digital twin concept is broadly adopted among manufacturing and services industries. Digital Twin technology is responsible for transforming business across sectors. Different domains, like agriculture, automobile, healthcare, construction, aeronautical, oil, gas, etc., are now adopting digital twin. The chapter discusses the application of digital twin among various industries. This chapter gives an overview of the digital twin to build an efficient supply chain. Based on the applications, the chapter identifies 15 benefits of a digital twin to build an efficient supply chain. Further, the chapter also reveals a few challenges in adopting digital twin.

Keywords Digital twin · Supply chain

1 Introduction

The supply chain complexities are increasing in today's fast-changing and challenging environment. To minimize it, organizations are now focusing on the adoption of new technologies. Industry 4.0 technologies are gaining massive popularity in various industries. One of the Industry 4.0 technology is digital twin technology. The concept of the digital twin is gaining close attention from many academicians and practitioners [75, 80], yet Gerlach et al. [26] and Cimino et al. [15] revealed that there is a lack of shared understanding of digital supply chain twin in the scientific literature. Marmolejo-Saucedo et al. [44] highlighted that a digital twin is a digital image or digital representation of anything. According to Grand View Research [28], the global digital twin market size was valued at USD 7.48 billion in 2021.

L. P. Vishwakarma (✉) · R. K. Singh
Management Development Institute, Gurgaon, India
e-mail: fpm21laxmi_p@mdi.ac.in

R. K. Singh
e-mail: rajesh.singh@mdi.ac.in

K E K Vimal et al. (eds.), *Industry 4.0 Technologies: Sustainable Manufacturing Supply Chains*, Environmental Footprints and Eco-design of Products and Processes,
https://doi.org/10.1007/978-981-99-4819-2_8

This market size is now projected to grow at a 39.1% Compound Annual Growth Rate (CAGR) (to reach USD 155,839.4 million) from 2022 to 2030. MarketsandMarkets [43] predicted that the digital twin market will grow at a CAGR of 60.6% (USD 73.5 billion) from 2022 to 2027. In a study by Fortune Business Insights [22], the digital twin market is forecasted to grow at a 40.6% CAGR (to reach USD 96.49 billion) by 2029.

The primary reasons behind this increase in the global digital twin are:

- The COVID-19 pandemic has speeded up the adoption of digital twin technology across different sectors such as manufacturing, health care, retail, oil and gas, telecommunication, automobile, real estate, and aerospace [28, 43, 47].
- A significant increase in the vendors offering digital twin technology, such as Oracle, Siemens, Hitachi Ltd., IBS, GE, SAP, Amazon web services, and Honeywell [43, 47].
- Increasing integration of Industry 4.0 technologies such as Cyber-Physical Systems (CPS), Internet of Things (IoT), Artificial Intelligence (AI), Big Data Analytics (BDA), and cloud computing with digital twin technology [22, 28].

Many big giants have aimed to deploy digital twin solutions for their supply chain. For example, Google Cloud and Microsoft Azure have launched multiple cloud-based digital solutions for industries for customized solutions and easy accessibility. In Jan 2022, Google cloud introduced a digital supply chain twin to offer unprecedented visibility of operations across the manufacturers and distributors in the supply chain. Such innovative solutions have triggered the adoption of a digital twin across industries [28]. Furthermore, digital twins are utilized by several leading organizations such as Oracle, NASA, Altair, General Electric, and US Airforce [46, 49, 62]. DHL Trend Research [16] revealed that the underlying digital technologies of digital twins include the IoT, AI, cloud computing; augmented, mixed, and virtual reality; application programming interface, and open standards. Dubey et al. [17, 18] revealed that digital technologies influence adaptability, agility, and alignment's impact on supply chain performance. Deployment of a digital twin enables the monitoring and collaboration of businesses with multiple partners. A digital supply chain twin helps understand the behavior of any supply chain, which further helps monitor the risks, test contingencies, optimize inventory and streamline the business operations. In a supply chain, the digital twin helps assess the demand and supply changes and helps optimize transportation resources by using real-time data. It provides end-to-end visibility in the supply chain, which helps in eliminating bottlenecks and identifying weaknesses [13, 26, 35, 36, 61, 63, 67, 69, 70, 73].

In a study by Forbes Technology Council, Koshulko [41] highlighted that medium size businesses will benefit the most from adopting digital twins. It is because the supply chain of the medium size firms is usually too complex to manage manually or with the help of any software. Complex supply chains can be handled easily using digital twins. Therefore, digital twins have become a significant trend in the logistics and supply chain management literature. A digital twin not only simulates the present state of the supply chain processes but also highlights what the future may look like depending on the number of factors and constraints [42]. Van der Valk et al. [75]

developed the taxonomy of digital twins. Barykin et al. [8] highlighted the role of the digital twin in supply chain and production management. The study reveals that the digital twin is considered the optimization procedure of industry 4.0. Barykin et al. [9] and Ivanov et al. [37] discussed the concept of the supply chain digital twin. The study highlights that combining optimization, simulation, and data analytics are the essential technology elements for creating any supply chain digital twin model. However, digital twins' production process includes big data, simulation, and the internet of things.

Given the wide applications of digital twin, it is important to uncover the opportunities and challenges in the context of supply chain. Despite getting a considerable number of articles examining the adoption of digital twin, it is difficult to find articles discussing opportunities and challenges of adopting digital twin for efficient supply chain in a single study. Our study tries to bridge this gap. To the best of our knowledge, this study makes a three-fold contribution. First, it provides the applications of digital twin across multiple industries for an efficient supply chain. Second, it focuses on the benefits of digital twin applications for an efficient supply chain. Third, the chapter examines the challenges for the digital twin adoption. All three contributions are important to have a better understanding of the adoption of digital twin. The following chapter is structured as follows. Section 2 discusses the definitions of digital twin. Section 3 illustrates the applications of a digital twin for an efficient supply chain across different industries. Section 4 examines the benefits of digital twin applications for an efficient supply chain. Section 5 focuses on the challenges for the digital twin adoption. Finally, Sect. 6 provides the conclusion for this chapter.

2 Definitions of Digital Twin

The increasing complexities and changing dynamics of supply chains are now creating significant challenges. The ongoing digital transformation of supply chains is a part of these challenges, but rather it provides an enormous opportunity and advantages compared with the traditional supply chain. The need for real-time data, unpredictable disruptions, and supply chain complexities changes the conventional supply chain and influences the adoption of Industry 4.0 technologies across the supply chain. Production wastes and unplanned downtime have negatively affected several manufacturers. Therefore, developing a system that can reduce the cost and spare time of manufacturing the products is essential. The demand to adopt an agile approach and analytical capabilities accelerated the need for a digital twin [25]. A prototype is created and used to run simulations by implementing digital twins, saving time, and costs. The data used for simulations can be both historical and real-time data, providing a more prominent and precise picture of the supply chain behavior. Many firms use the digital twin to increase process efficiency, analyze trends, monitor risks, and simulate any changes they want to implement soon. Digital twins help firms solve problems faster and uncover potential flaws in all phases [43].

There is no uniform understanding of the digital supply chain twin [75]. But past literature has defined the digital twin. For instance, few definitions are as follows:

Barnard [6] defines the digital twin as "*A digital twin is a model which describes how an actual original item—a machine, component, process,* etc.—*behaves. Mathematical functions describe how the relevant variables—input values, physical parameters (e.g., friction or temperature), and output values, for example—depend on each other. Such a function could determine, for example, that the temperature in a motor increases by one degree for each minute it is in operation.*"

Miller et al. [45] defined the digital twin as "*A Digital Twin is a digital representation of a physical thing's data, state, relationships, and behavior.*"

Miller et al. [45] further discussed the six major characteristics of a digital twin. These characteristics are:

i. Physical and virtual systems: Both the material and the virtual system are considered as a part of digital twin.
ii. Bidirectional data: A digital twin supports the data exchange in both directions between the physical and the virtual system.
iii. Maintain state: A digital twin should be able to save the last state of the physical system to avoid any delay with disconnections.
iv. Modeling and analysis: A digital twin should have modeling and analytical capabilities.
v. Timely updates: A digital twin should be able to provide well-timed updates based on the use case requirements.
vi. Reporting: A digital twin should pass the results to the machines or people.

Grieves and Vickers [29] observed the digital twin "*The Digital Twin is a set of virtual information constructs that fully describes a potential or actual physical manufactured product from the micro level to the macro geometrical level. Therefore, at its optimum, any information that could be obtained from inspecting a physical manufactured product can be obtained from its Digital Twin*".

Glaessgen and Stargel [27] explained the digital twin as "*A Digital Twin is an integrated multiphysics, multiscale, probabilistic simulations of an [...] system that uses the best available physical models, sensor updates, [...], to mirror the life of its corresponding [...] twin. The Digital Twin is ultra-realistic and may consider one or more important and interdependent [...] systems [...]. The Digital Twin integrates sensors data from the [...] system, [...] and all available historical [...] data obtained using data mining and text mining.*"

In addition, other authors have explained that the digital twin as: Digital twin is a conceptualized digital SC model replicating the network state in real-time data and surveillance systems [2, 24, 49]. A digital twin represents the physical SC based on inventory demand, transportation, and capacity data. Digital twins are used for planning and real-time control decisions.

Further, the definitions and understanding of digital supply chain twins are also provided by several authors.

Marmolejo-Saucedo et al. [44] illustrated digital supply chain twins as "*a detained simulation model of an actual supply chain which predicts the behavior and dynamics of a supply chain to make mid-term/short-term decisions. It comprises six layers: the physical twin, the local data source, local data repositories, IoT gateway interfaces, Cloud-based information repositories and the emulations and simulation platform.*"

Ivanov et al. [37] highlighted it as "*a model that can always represent the network state in real-time for any given moment and interacting with other supply chain management tools provides a control tower for complete end to end supply chain visibility to improve resilience and test contingency plans.*"

Korth et al. [40] examined as "*it is a linked collection of different types of data (like operations data) as well as different models; it evolves with the real system along its life cycle; it is able to derive solutions relevant for the real systems (e.g., optimize, operation and service).*"

3 Applications of a Digital Twin for an Efficient Supply Chain Across Different Industries

Several industries widely accept digital twins, such as automotive [13, 42, 58]; manufacturing [16, 56, 72]; material science [16]; smart city [56]; energy [16, 69]; life science and health care [16, 56]; agriculture [56]; and consumer, retail, and e-commerce [16]. Srai et al. [70] discussed digital twin applications in the organic food supply chain, pharmaceutical supply chain, and precision agriculture. The study further reveals the opportunities and challenges of using digital twin in these areas. In the organic food supply chain, digital twin helps in combining the digital and analytical technologies, while, for a pharmaceutical supply chain, digital twin helps in reducing the inventory, reconfiguring the information flow, process flow, and material flow. For a precision agriculture, digital twin helps in effective resource utilization and improving the farmer's livelihood. Further, the challenges faced across these fields include technology complexity, lack of technical infrastructure, lack of required skills, and lack of policies. Gerlach et al. [26] demonstrated that digital twins are used in the areas of warehouse management, manufacturing, logistics management, network management, cargo handling, transportation management, facility management, asset management, and shipment management [16, 31, 40, 65]. In these areas, digital twins help in risk management, transportation flexibility, shop floor management, sustainability assessments, production plannings, transportation planning, inventory management, reducing the bottlenecks, reducing lead time, building supply chains resilient, increasing supply chain efficiency, improving the service levels, providing supply chain transparency, cargo load planning, and indoor and outdoor vehicle dispatching. The challenges faced are changing customer requirements, changing market conditions, and lack of experts in the area of digital twins implementation.

3.1 Agri-Food

With the changing environmental conditions and consumer demands, agri-food supply chains are continuously changing. Digital twins are helping to resolve the challenges of an agri-food supply chain [61]. The challenges, such as limited shelf-life, curbing greenhouse gas emissions, and food losses, are minimized using digital twin technology. Yet, Tzachor et al. [73] observed that the capability of the digital twin in agricultural production is yet to be realized. For an Agri-food supply chain, the digital twin application starts at the farmer's level and ends at the consumer's level. A digital twin helps the farmers to predict the atmosphere conditions, soil temperature, irrigation, and nutrition required for the seeds. Based on the requirements, digital twin provides actionable recommendations. For example, digital twin offers recommendations for applying pesticides and land management optimally. The sensors used in the digital twin monitor the real-time environmental parameters such as the humidity level, air temperatures, and oxygen concentration required during farming. Depending on the product life cycle, the digital twin helps analyze the life cycle, manufacturing process, and operations of particular food items. Digital twins are used to structure the design and functions of the agri-food supply chain to optimally use light intensity, CO_2 concentrations, humidity, water nutrient recycling, and temperatures [73].

Further, Pylianidis et al. [61] revealed that digital twins perform real-time system failure prediction and analysis, technology integration, virtual maintenance, energy consumption analysis, monitoring, and optimizing operations in agriculture. Digital twins have wide application in cold chains of perishable products too. It monitors the perishable products' humidity levels, respiratory behavior, food temperature, delivery schedules, and grid carbon intensity. Digital twin, combined with other Industry 4.0 technologies, helps food processing and packaging facilities. The digital twin models help optimize the food distribution systems, reducing food wastage and maintaining food inventory in retail stores [73].

3.2 Aerospace and Aeronautical

American Institute of Aeronautics and Astronautics [3] defined the digital twin as "*[...] a set of virtual information constructs that mimics the structure, context, behavior of an individual unique physical asset, or a group of physical assets, is dynamically updated with data from its physical twin throughout its lifecycle and informs decisions that realize value.*" Digital twin supports the designing of products (such as pumps, engines, and turbines) and processes and contributes to the manufacturing process (including inspection of the product and processes, repairing and providing maintenance to the products). The digital twins are constructed differently for each product line and component [74]. For example, Glaessgen et al. [27] highlighted that NASA uses the digital twins' concept for manned and unmanned aircraft. Airservices Australia uses digital twin for air traffic management. The firm uses a digital twin to resolve air traffic issues. Digital twin helps the firm to reduce delays, improve network planning, enhance flight routes, and optimize takeoff times. It further helps the firm to respond quickly to changing customer requirements and requests. It enables data-driven decision-making and high-precision predictions [32, 54]. Qi and Tao [62] highlighted that digital twins are used to diagnose problems virtually and proactively maintain the aircraft structures. Tao et al. [72] revealed that the digital twin predicts the service life of aircraft by monitoring the aircraft's design. The study further highlights that digital twins are also used to monitor the operational state and analyze the wing and tires of aircraft structural damages.

3.3 Automobile

In the automobile industry, the application of digital twin is broadly observed in the planning, manufacturing, designing, and maintenance departments. Fraga-Lamas and Fernández-Caramés [23] revealed that the digital twin of a vehicle provides the vehicle's past and future state. Digital twin provides the virtual model of the cars even before it is constructed or produced. For example, Mercedes-AMG Petronas Formula One uses digital twins to create virtual models of vehicles and racing tracks. Data is the main element of the simulation work. The simulation mimics real-time experiences, which helps the firm accelerate the designing, testing, and manufacturing of improved car parts for the next race. The simulation involves data science, interactive visual analytics, setup parameters, and "what if" scenarios to optimize the functional areas. The simulation increases the team's ability to speed up and fine-tune themselves during the seasons. Digital twin helps to identify weak points and improvement areas for improving performance. It further helps to add new design features. Overall, the digital twin helps to gain a competitive advantage by getting deeper insights for operational optimization and producing better results [57].

Furthermore, NASA uses digital twins to develop vehicles for extreme conditions [76]. Tao et al. [71] observed that digital twins are used to analyze the oil pressure, engine speed, and other essential parameters to prevent vehicle breakdown. Schleich et al. [66] highlighted that digital twins are used to maintain automobile parts like oil replacements. In another example, Parrott et al. [54] assessed that Bridgestone uses digital twin simulations augmented by sensor data to improve the performance of tires. In addition, the company uses the digital twin to get insights across the supply chain.

3.4 Health Care

A digital twin has tremendous potential to change the working environment of the healthcare industry. Newman [50] argued that the true potential of the digital twin in health care is mainly unexplored. Elkefi and Asan [19] revealed that the digital twin holds the capability to strengthen the overall healthcare processes. Its application is for different patients. Every patient is different; therefore, every corresponding digital twin model is different throughout the patient's lifetime. A digital twin is practical when a patient's data (physiology of the patient or specific anatomical features) is incorporated. For example, a specific CAD geometry was obtained from a particular patient's MRI or X-ray report to build a digital twin model [76]. Digital twins help evaluate the "what-if" scenarios in diagnostic testing for personalized therapy. The "what-if" scenarios help reorganize the clinical and patient's workflow in particular departments [68]. The simulations also help in splitting and re-arranging waiting areas. It further helps perform operational stress testing for a specific department or hospital. Digital twin provides the hospital's real-time operations, which helps optimize resources [68].

Furthermore, digital twins have broad applications in designing medical equipment, improving health outcomes, performing surgery, improving the healthcare stakeholders' (patients and doctors) experiences at minimal risk, precision medicine, drug treatments, and therapy [19, 21]. For example, Philips uses a digital twin to design medical equipment for personal health and hospital requirements. The firm conducted a simulation to develop a portable oxygen generator for patients with breathing problems. The firm uses digital prototypes to test the performance of portable oxygen generators and performs several iterations until the prototype is developed [76]. Another example is Mater private hospital, a leading cardiac care institution and radiologic imaging institution in Dublin. The hospital has identified several improvements after adopting digital twin. The hospital was able to increase the equipment utilization (CT scan usage increased by 26% and MRI usage increased by 32%); faster the patient's turnaround time (reduced by 28 min for CT scan and 34 min for MRI); shorter patient's wait time (13 min for CT scan and 25 min for MRI) and lowers the staffing cost. Overall, the hospital improved the patient's experience and utilized the hospital space [68]. The study further highlights that a digital twin is used for cardiology operations, prostate cancer treatment, and intensive care units. In addition, the study reveals that digital twins are used to design and provide better treatment for many disease-like cardiovascular diseases.

3.5 Construction

The increasing complexity of the construction industry gives rise to the adoption of digital twin [38]. Brilakis et al. [12] revealed that the digital twin is considered a critical enabler that can enhance the construction industry's poor digit history. Ottinger et al. [51] highlighted that the adoption of the digital twin in the construction and real estate industry could lead to a 35% increase in operational efficiency and building maintenance, 20% increase in productivity, 50% increase in reduction the greenhouse emission and carbon footprints, 50% increase in resilience, and 15% increase in space utilization. But Ammar et al. [4] highlighted a lack of a universal definition of digital twins in the construction industry. Khajavi et al. [39] highlight digital twins are used in lean construction, clash detection, efficient consumption of resources, stakeholder interoperability, closed-loop designs, design visualization, enhanced user comfort, cost, and time estimation. Yitmen et al. [79] revealed that digital twins are used to monitor, diagnose, and optimize asset conditions. In addition, digital twins further enable preventive predictions whenever required.

The applications of the digital twin in the construction industry are to enhance the health and safety, proficiency of the workforce, improve workforce productivity, efficient facility management, and minimize the cost of operations and construction [1, 51]. For example, Plautz [59] highlighted that Las Vegas unveiled a digital twin of 7 km. As a result, the city has taken action to mitigate the challenges such as traffic management, optimally energy usage, mobility issues, air quality, noise, parking, emission, and emergency management. The digital twin model uses the real-time data from the 5 g network and the Internet of Things sensors to visualize, plan and train the project organizers (architects, real estate owners, and casino operators) accordingly.

3.6 *Oil and Gas*

A digital twin can leverage the oil and gas industry in multiple ways. Broadly, it can be incorporated into the production system, subsea, umbilicals, risers, and flowlines. Considering the uncertain market scenarios, such applications attract oil and gas firms to adopt digital twin technology. A digital twin enables automatic improvements (for example, To alter valve settings—a particular algorithm can be used). Digital twin saves around 9–15% of the total decommissioning project cost [48]. Holmås et al. [34] revealed that "*A digital twin could provide an operator with a bird's—eye view of flows throughout a pipeline.*" Digital twin helps create and predict the "what-if" model for different scenarios. Such scenarios reduce the time to estimate the project parameters, conduct dry runs to optimize the asset value, speed up the testing alternatives, and improve the trade-off between capital and operational expenditure. For example, British Petroleum's digital twin infrastructure APEX is a developed simulation and surveillance system that creates the production system's virtual models. Having digital twins installed at 11 refineries enhances the firm's process optimization and operational gains while saving $154 million [48].

4 Benefits of Digital Twin Applications for an Efficient Supply Chain

The various benefits of a digital twin for an efficient supply chain are discussed below.

4.1 Supply Chain Planning

Supply chain planning is the foundation for sourcing, production, and logistics. Supply chain planning is crucial as it directly impacts the overall quality, cost, and productivity [77]. Digital supply chain twins positively contribute to planning multiple areas of the supply chain, like production planning, maintenance planning, transportation planning, sales, and operations. Digital supply chain twins, AI algorithms, and other technologies identify the system constraints and suggest maintenance plans for smooth future works. The timely maintenance of machines in the supply chains helps to manufacture additional capacities whenever needed. Gerlach et al. [26] highlighted that a digital twin guides the changing dynamics of transportation processes and contributes to transportation planning. It operationalizes and synchronizes the system based on real-time data. It stimulates the near future scenarios and estimates what would be the congestion levels. Based on the simulation results, it suggests to the decision-makers if any freight re-route is required. Digital supply chain twins help optimize sales and operations planning by simulating the execution plan, highlighting the risks and opportunities, and providing feedback to the planning processes. It allows the firms to minimize the losses and the future bottlenecks that may arise from the misalignment of plans and system constraints [67].

4.2 Managing Risks and Building Supply Chain Resilient

The disruption directly or indirectly hinders the supply chain performance by disturbing the supply chain. It occurs due to low inventory, stocks return, firm's competitive positioning in the market. The disruptions made the supply chain collapses and resulted in causing ripple effects. Such effects cause severity and stress to the global supply chain, and the digital twin identifies where disruptions and volatility exist before the issue arises [37]. Gerlach et al. [26] highlighted that digital supply chain twins identify the risks early and provide means to mitigate those risks. The simulation models in the digital twins take the data about the business processes, facility locations, customer demand, inventory policies, and production capabilities. The collected data are updated in real time, which helps build "what-if" scenarios to identify the risks and therefore helps build a resilient supply chain in the long run. Barykin et al. [9] revealed a supply chain digital twin is more reliable during any uncertainty. Modern technologies in the digital twin collect a large amount of data in the supply chain to identify production capabilities, financial conditions, and supply failure. The critical hotspots can easily be found using a digital twin and help provide early warning signals about the incidents that might happen and disturb the supply chain. It further helps identify political, financial, or natural risks.

4.3 Mapping the Supply Chain Network and Ensuring Visibility

Digital twin helps in understanding the supply chain dynamics and behavior. A digital twin maps the supply chain network and ensures visibility [14]. Koshulko [41] encores that digital supply chain twins analyze the risks, inventory levels, KPIs, demand levels, and supplier and sales data. Therefore, it can ensure the optimal course of action in any circumstances. Hence, it provides visibility across the supply chain. For example, the digital twin models picture the state of working on the shop floor using robots or machines. The model helps to attend the shop floor meetings without being physically present [11]. For example, Intermarché, a French supermarket chain, uses the internet of things to enable sales and shelves systems to develop a digital twin of stores. The model provides real-time monitoring of stores. It improves the efficiency of each store, maps the supply chain network, and ensures visibility [47].

4.4 Analyzing Demand and Supply Signals

Digital supply chain twins enable the firm to reduce the idle time of the bottleneck, enhance the inventory positioning, and analyze the demand and supply signals [67]. Furthermore, managing risks and analyzing demand and supply signals helps the firm to speed up its reaction time according to the market needs [37].

4.5 Testing the Contingency Plans

The digital twin simulation models help test the supply chain contingency plans. Based on the results, the model provides suggestions to improve and make the necessary changes [10, 37]. Testing the contingency plans saves the firm's time, cost, resources, and energy [63].

4.6 Accurate Forecasting and Testing Operations

The digital twin predicts future impacts on a supply chain [64]. It helps forecast and test the operations over the coming weeks and months [10]. Wang et al. [77] highlight that forecast is the base for any supply chain planning. A digital twin supply chain helps forecast demand and determines the required schedules, investments, and products [77].

4.7 Predictive Maintenance

Digital twins now enable predictive maintenance [65]. Rather than guessing when the machines or the products need service, the sensors help check the irregularities quickly. It further helps in determining the wear and tear of devices or products. It helps in avoiding unexpected and expensive repairs. Mussomeli et al. [47] revealed that digital twins identify equipment breakdown time and various engine components' life spans and estimate maintenance and non-maintenance procedures. Park et al. [53] highlighted that digital twin models adjust according to dynamic instabilities like the bullwhip effect, efficiently controlling the supply chain and reducing downtime.

4.8 Boosting Innovations

Van Houten [76] encores that adopting a digital twin across several industries boosts innovations. The primary reason behind this is that a digital twin enables the rapid prototyping of any new product, thereby leading to product development. The virtual representation saves costs and provides the final product's idea (if any changes are required). Mussomeli et al. [47] revealed the simulation of new ideas and prototyping of new products boost innovations.

4.9 Cost Savings

A digital twin helps relax the transportation constraints, capacity constraints, supplier and producer constraints simulates the supplier issues, and changes the region of supplier or customer to optimize costs [63]. Furthermore, with the allocation of customers, digital twin helps save operational costs [68].

4.10 Contributes to Sustainability Assessments

Barykin et al. [9] highlight that a digital supply chain twin positively contributes to sustainability assessments. Pehlken and Baumann [55] highlight that digital twins aim to identify recycling potential, reduce material waste, decrease energy consumption, and improve material efficiency and ecological sustainability. Barni et al. [7] revealed that digital twin models assess different scenarios by changing the production techniques, demand fulfillment rates, and modes of transportation and adjusting the decision-making in terms of sustainability.

4.11 Optimizes and Supports Strategic Decisions

Marmolejo-Saucedo et al. [44] encored that a digital twin compares and aligns the supply chain planning. This planning optimizes and supports the strategic decision across the vertical dimension (connecting the strategy to execution) and the horizontal size (all along the supply chain). It can be used to build a model that either focuses on one functional area (like warehousing or logistics) or the end-to-end model [63]. These models help in both the short-term and the long-term decision-making processes. For example, a model predicting the short-term stock-outs of the supply chain helps make short-term decisions.

In contrast, a model helps optimize the transportation network and helps in making long-term decisions. Digital twin enables scenario planning based on the firm's need, which allows the firm to make crucial decisions. Supply chain digital twins make recommendations based on "what-if" scenarios by analyzing vertically and the horizontally throughout the supply chain. These "what-if" scenarios are run in the digital twin models, encapsulating the suppliers, manufacturers, transportation, distributors, retailers, and the customer's location details. Digital twin optimizes a supply chain's strategic and logistical locational decisions [10, 64].

4.12 Accelerating Product and Process Developments

Sanderse and Weippl [65] observed that digital twins accelerated product and process developments. Further, Pan et al. [52] and Hofmann and Branding [33] observed that digital twin models improve process efficiency. For example, in cargo handling, digital twin models consider load requirements and, based on them, position the load efficiency. Such a process optimizes and supports the entire supply chain process [78]. Srai et al. [70] highlight that in the pharmaceutical industry, most of the digital twin developments lie in combining experimentation and model-based evidence to predict the new product properties and understand the manufacturing technology required for the new product.

4.13 Inventory Management

Digital supply chain twin models simulate and estimate the inventory required shortly. It then tests various transportation and inventory policies based on the inventory requirements. Further, the simulation experiments also measure inventory's effect on the logistics system's performance [26]. Semenov et al. [67] observed that digital twin helps eliminate bottlenecks, improve service levels, reduce lead time, and provide transparency in inventory channels. Wang et al. [77] observed that a digital twin helps engineers to verify inventory planning and distribution. It improves the inventory cycle, inventory cycle costs, safety inventory levels, and seasonal inventory levels.

4.14 Enhancing the Product and Process Quality

Wang et al. [77] highlighted that the digital twin enhances the process quality of virtual modeling by simulation results. Virtual simulations polish the supply chain, promote high-quality models, perform more iterative simulations, take more abundant data, and improve the process quality.

4.15 Improving Supply Chain Performances

Wang et al. [77] encored that a digital twin performs faster action to reduce the lead time. The supply chain planning, analysis of demand and supply signals, testing the contingency plans, forecasting, predictive maintenance, innovations,

Table 1 Benefits of digital twin in the supply chain

Benefits of digital twin	References
Supply chain planning	Gerlach et al. [26]; Semenov et al. [67]
Managing risks and building supply chain resilient	Barykin et al. [9]; Ivanov et al. [37]
Mapping the supply chain network and ensuring visibility	Koshulko [41]; Ivanov and Dolgui [36]; Choi et al. [14]; Ivanov [35]; Mussomeli et al. [47]
Analyzing demand and supply signals	Semenov et al. [67]; Ivanov et al. [37]
Test the contingency plans	Quintanilla [63]; Ivanov et al. [37]; Battarra et al. [10]; Salman and Yücel [64]
Accurate forecasting and testing operations	Wang et al. [77]; Battarra et al. [10]; Salman and Yücel [64]
Boosting innovations	Mussomeli et al. [47]; van Houten [76]
Predictive maintenance	Park et al. [53]; Sanderse and Weippl [65]; Mussomeli et al. [47]
Cost savings	Quintanilla [63]; Siemens Healthineers [68]
Contributes to sustainability assessments	Pehlken and Baumann [55]; Battarra et al. [10]; Barni et al. [7]
Optimizes and supports strategic decisions	Quintanilla [63]; Marmolejo-Saucedo et al. [44]; Battarra et al. [10]; Salman and Yücel [64]
Accelerating product and process developments	Pan et al. [52]; Wong et al. [78]; Eschemann et al. [20]; Ashrafian et al. [5]; Hofmann and Branding [33]; Srai et al. [70]; Sanderse and Weippl [65]
Inventory management	Gerlach et al. [26]; Semenov et al. [67]
Enhancing the product and process quality	Wang et al. [77]; Semenov et al. [67]
Improving supply chain performances	Quintanilla [63]; Pan et al. [52]; Gerlach et al. [26]; Eschemann et al. [20]; Wang et al. [77]

product, process developments, and improving the overall supply chain performance [20, 52, 63].

Table 1 summarizes the benefits of digital twin in the supply chain.

5 Challenges of Adopting a Digital Twin

The adoption of a digital twin supply chain comes with a few challenges. First, data security and privacy are the primary concern for any organization implementing digital twin, raising the problem of cyberattacks [19, 60]. Second, cloud-based digital twins save information online, increasing security and privacy concerns. Third, small enterprises found it very difficult to invest in digital supply chain twins due to a lack of investments, government incentives, and expected cyber-attack challenges [16, 73]. Fourth, the lack of skilled professionals makes handling a large amount of data and developing strategies against cyber-attacks challenging. Finally, poor data quality further hinders the adoption of digital twins [16, 60].

Elkefi and Asan [19] revealed that the infrastructure required for data flow is missing for implementing the digital twin. Popa et al. [60] highlighted that adopting digital twins creates socio-ethical risks such as sharing personal content, privacy issues, and disruption of existing societal structures. The study highlights that the adoption of digital twins creates inequality and injustice in majorly two forms. First, this technology increases the socio-economical gap because it is not accessed by everyone and does not cover health insurance for everyone. Second, this technology is not yet adopted by the southern and eastern countries, thereby increasing the gap between the rich and the poor. Marmolejo-Saucedo et al. [44] examined that information technology integration, information rights, digital security, and integration of partner companies are significant challenges in implementing digital twins in any firm. DHL Trend Research [16] observed that the difficulties of applying digital twins are specific representation issues, interoperability issues, and IP protection risks.

6 Conclusion

The supply chain is the most crucial part of any industry. Industry 4.0 technologies are now contributing hugely to the development and improvement of supply chains. One of those techniques is the digital twin, the virtual models highlighting the supply chain's working. These virtual models are well used in different industries to make an efficient supply chain. A few of those industries are Oracle, NASA, Altair, General Electric, and US Airforce [46, 62]. Digital twin provides an opportunity to accelerate innovation among multiple industries. It changes the traditional system and imputes the digital orientation of processes, products, and services [30]. This chapter discusses the applications of digital twin among the agri-food, aerospace and aeronautical, automobile, health care, construction, oil and gas industries.

In addition, the study identifies numerous benefits of digital twin applications to make an efficient supply chain. The application of digital twin provides benefits such as supply chain planning, managing risks, building supply chain resilient, mapping the supply chain network and ensuring visibility, analyzing demand and supply signals, deploying recovery policies, testing the contingency plans, accurate forecasting, and testing operations, predictive maintenance, boosting innovations, cost savings; contributes to sustainability assessments, optimizes and supports strategic decisions, accelerating product and process developments inventory management, enhancing the product quality, and improving the customer services and supply chain performances. Our study's findings will motivate firms that have not yet adopted the digital twin for their supply chains. The results of this study are drawn on the basis of a review of relevant studies conducted by academicians and practitioners. This chapter discusses multiple examples of industries and firms adopting digital twins to build an efficient supply chain. This study provides an understanding of digital twins among diverse sectors and how the firms benefit from using digital twins. More empirical studies can be conducted in the future to explore the benefits discussed in this chapter.

References

1. Akanmu AA, Anumba CJ, Ogunseiju OO (2021) Towards next generation cyber-physical systems and digital twins for construction. J Inf Technol Constr 26:505–525
2. Alla AA, Kreutz M, Rippel D, Lutjen M, Freitag M (2019) Simulation-based analysis of the interaction of a physical and a digital twin in a cyber-physical production system. IFACPapersOnLine 52(13):1331–1336. https://doi.org/10.1016/j.ifacol.2019.11.383
3. American Institute of Aeronautics and Astronautics (2020) Digital Engineering Integration Committee. Digital twin: definition and value. AIAA and AIA Position Paper.
4. Ammar A, Nassereddine H, Abdulbaky N, Aboukansour A, Tannoury J, Urban H, Schranz C (2022) Digital twins in the construction industry: a perspective of practitioners and building authority. Front Built Environ 102
5. Ashrafian A, Pettersen OG, Kuntze KN, Franke J, Alfnes E, Henriksen KF, Spone J (2019, September) Full-scale discrete event simulation of an automated modular conveyor system for warehouse logistics. In: IFIP international conference on advances in production management systems. Springer, Cham, pp 35–42
6. Barnard A (2021) Siemens. Digital twins: faster, easier, and reusable. https://new.siemens.com/global/en/company/stories/research-technologies/digitaltwin/x-digital-twin.html. Accessed 7 Sept 2022
7. Barni A, Fontana A, Menato S, Sorlini M, Canetta L (2018) Exploiting the digital twin in the assessment and optimization of sustainability performances. In: Proceedings of the 2018 international conference on intelligent systems (IS), Funchal, Portugal, 25–27 September 2018, pp 706–713
8. Barykin SY, Bochkarev AA, Dobronravin E, Sergeev SM (2021) The place and role of digital twin in supply chain management. Acad Strateg Manag J 20:1–19
9. Barykin SY, Bochkarev AA, Kalinina OV, Yadykin VK (2020) Concept for a supply chain digital twin. Int J Math, Eng Manag Sci 5(6):1498
10. Battarra M, Balcik B, Xu H (2018) Disaster preparedness using risk-assessment methods from earthquake engineering. Eur J Oper Res 269(2):423–435. https://doi.org/10.1016/j.ejor.2018.02.014

11. Brenner B, Hummel V (2017) Digital twin as enabler for an innovative digital shop floor management system in the ESB Logistics Learning Factory at Reutlingen-University. Procedia Manuf 9:198–205
12. Brilakis I, Pan Y, Borrmann A, Mayer H-G, Rhein F, Vos C, et al (2019) Built environment digital twinning. In: International workshop on built environment digital twinning presented by TUM Institute for Advanced Study and Siemens AG. https://publications.cms.bgu.tum.de/reports/2020_Brilakis_BuiltEnvDT.pdf. Accessed 5 Nov 2021
13. Caputo F, Greco A, Fera M, Macchiaroli R (2019) Digital twins to enhance the integration of ergonomics in the workplace design. Int J Ind Ergon 71:20–31
14. Choi TY, Rogers D, Vakil B (2020) Coronavirus is a wake-up call for supply chain management. Harv Bus Rev
15. Cimino C, Negri E, Fumagalli L (2019) Review of digital twin applications in manufacturing. Comput Ind 113:103130
16. DHL Trend Research (2022) Digital twin in logistics. https://www.dhl.com/content/dam/dhl/global/core/documents/pdf/glo-core-digital-twins-in-logistics.pdf. Accessed 6 Sept 2022
17. Dubey R, Altay N, Gunasekaran A, Blome C, Papadopoulos T, Childe SJ (2018) Supply chain agility, adaptability and alignment: empirical evidence from the Indian auto components industry. Int J Oper Prod Manag
18. Dubey R, Gunasekaran A, Childe SJ, Fosso S, Roubaud D, Foropon C (2021) Empirical investigation of data analytics capability and organizational flexibility as complements to supply chain resilience. Int J Prod Res 59(1):110–128
19. Elkefi S, Asan O (2022) Digital twins for managing health care systems: rapid literature review. J Med Internet Res 24(8):e37641
20. Eschemann P, Borchers P, Feeken L, Stierand I, Zernickel JS, Neumann M (2020) Towards digital twins for optimizing the factory of the future. Proceedings of the Modelling and Simulation.
21. Feng Y, Zhao J, Kleinstreuer C, Wang Q, Wang J, Wu DH, Lin J (2018) An in silico inter-subject variability study of extra-thoracic morphology effects on inhaled particle transport and deposition. J Aerosol Sci 123:185–207
22. Fortune Business Insights (2022) https://www.fortunebusinessinsights.com/digital-twin-market-106246. Accessed 2 Sept 2022
23. Fraga-Lamas P, Fernández-Caramés TM (2019) A review on blockchain technologies for an advanced and cyber-resilient automotive industry. IEEE Access 7:17578–17598
24. Frank AG, Dalenogare LS, Ayala NF (2019) Industry 4.0 on procurement and supply management: a conceptual and qualitative analysis. Int J Prod Econ 210:15–26
25. Galea-Pace S (2022) The evolution of digital twins in supply chain. https://supplychaindigital.com/technology/evolution-digital-twins-supply-chain. Accessed 2 Sept 2022
26. Gerlach B, Zarnitz S, Nitsche B, Straube F (2021) Digital supply chain twins—conceptual clarification, use cases and benefits. Logistics 5(4):86
27. Glaessgen E, Stargel D (2012, April) The digital twin paradigm for future NASA and US Air Force vehicles. In: 53rd AIAA/ASME/ASCE/AHS/ASC structures, structural dynamics and materials conference 20th AIAA/ASME/AHS adaptive structures conference 14th AIAA, p 1818
28. Grand View Research (2022) Digital twin market. https://www.grandviewresearch.com/industry-analysis/digital-twin-market. Accessed 2 Sept 2022
29. Grieves M, Vickers J (2017) Digital twin: mitigating unpredictable, undesirable emergent behavior in complex systems. In: Transdisciplinary perspectives on complex systems. Springer, Cham, pp 85–113
30. Guo J, Lv Z (2022) Application of digital twins in multiple fields. Multimed Tools Appl 1–27
31. Haße H, Li B, Weißenberg N, Cirullies J, Otto B (2019) Digital twin for real-time data processing in logistics. In: Artificial intelligence and digital transformation in supply chain management: innovative approaches for supply chains. Proceedings of the Hamburg international conference of logistics (HICL), vol 27. Epubli GmbH, Berlin, pp 4–28

32. Hendry J (2022) Airservices eyes fully-fledged digital replica of Australian airspace. ITnews. https://www.itnews.com.au/news/airservices-eyes-fully-fledged-digital-replica-of-aussie-airspace-576891. Accessed 8 Sept 2022
33. Hofmann W, Branding F (2019) Implementation of an IoT-and cloud-based digital twin for real-time decision support in port operations. IFAC-PapersOnLine 52(13):2104–2109
34. Holmås H, Sjåtil OA, Santamarta S, Lindseth S, Forbes P, Romanin P (2019) Creating value with digital twins in oil and gas. Boston Consulting Group, Boston, MA, USA
35. Ivanov D (2020) Predicting the impacts of epidemic outbreaks on global supply chains: a simulation-based analysis on the coronavirus outbreak (COVID-19/SARS-CoV-2) case. Transp Res Part E, Logist Transp Rev 136:101922. https://doi.org/10.1016/j.tre.2020.101922
36. Ivanov D, Dolgui A (2020) Viability of intertwined supply networks: extending the supply chain resilience angles towards survivability. A position paper motivated by COVID-19 outbreak. Int J Prod Res 58(10):2904–2915. https://doi.org/10.1080/00207543.2020.1750727
37. Ivanov D, Dolgui A, Das A, Sokolov B (2019) Digital supply chain twins: managing the ripple effect, resilience, and disruption risks by data-driven optimization, simulation, and visibility. In: Ivanov D, Dolgui A, Sokolov B (eds) Handbook of ripple effects in the supply chain. Springer, Cham, Switzerland, pp 309–332. ISBN 978-3-030-14301-5
38. Kan C, Anumba CJ (2019) Digital twins as the next phase of cyber-physical systems in construction. Computing in civil engineering 2019: data, sensing, and analytics. American Society of Civil Engineers, Reston, VA, pp 256–264
39. Khajavi SH, Motlagh NH, Jaribion A, Werner LC, Holmström J (2019) Digital twin: vision, benefits, boundaries, and creation for buildings. IEEE Access 7:147406–147419
40. Korth B, Schwede C, Zajac M (2018, December) Simulation-ready digital twin for real-time management of logistics systems. In: 2018 IEEE international conference on big data (big data). IEEE, pp 4194–4201
41. Koshulko A (2022) Forbes technology council. How digital twin can help supply chain survive disruption. https://www.forbes.com/sites/forbestechcouncil/2022/06/21/how-digital-twins-can-help-supply-chains-survive-disruption/?sh=36664f0a6800. Accessed 2 Sept 2022
42. Kritzinger W, Karner M, Traar G, Henjes J, Sihn W (2018) Digital twin in manufacturing: a categorical literature review and classification. IFAC-PapersOnLine 51(11):1016–1022
43. MarketsandMarkets (2022) Digital twin market. https://www.marketsandmarkets.com/Market-Reports/digital-twin-market-225269522.html. Accessed 2 Sept 2022
44. Marmolejo-Saucedo JA, Hurtado-Hernandez M, Suarez-Valdes R (2019, October) Digital twins in supply chain management: a brief literature review. In: International conference on intelligent computing and optimization. Springer, Cham, pp 653–661
45. Miller P (2020) Grasp the challenge of implementing digital twins at scale: digital twins show great promise, but early adopters struggle with technical and organizational barriers as they scale. https://www.forrester.com/report/Grasp-The-Challenge-Of-Implementing-Digital-Twins-At-Scale/RES158396. Accessed 2 Sept 2022
46. Mukherjee T, DebRoy T (2019) A digital twin for rapid qualification of 3D printed metallic components. Appl Mater Today 14:59–65
47. Mussomeli A, Meeker B, Shepley S, Schatsky D (2018) Expecting digital twins. Deloitte Insights. https://www2.deloitte.com/content/dam/insights/us/articles/3773_Expecting-digital-twins/DI_Expecting-digital-twins.pdf. Accessed 2 Sept 2022
48. Nair S (2020) The increasing popularity of digital twins in oil and gas. GEP Insight Drives Innovation. https://www.gep.com/blog/mind/the-increasing-popularity-of-digital-twins-in-oil-and-gas#:~:text=A%20digital%20twin%20is%20a,into%20performance%20and%20potential%20problems. Accessed 10 Sept 2022
49. Negri E, Fumagalli L, Macchi M (2017) A review of the roles of digital twin in CPS-based production systems. Procedia Manuf 11:939–948
50. Newman D (2019) Top 6 digital transformation trends in healthcare for 2019. Retrieved June 11, 2019, from Forbes website https://www.forbes.com/sites/danielnewman/2019/01/03/top-6-digital-transformation-trends-in-healthcare-for2019/amp/

51. Ottinger NB, Jordan Stein E, Crandon MG, Jain A (2021) Digital twin: the Age of Aquarius in construction and real estate
52. Pan YH, Wu NQ, Qu T, Li PZ, Zhang K, Guo HF (2021) Digital-twin-driven production logistics synchronization system for vehicle routing problems with pick-up and delivery in industrial park. Int J Comput Integr Manuf 34(7–8):814–828
53. Park KT, Son YH, Noh SD (2021) The architectural framework of a cyber physical logistics system for digital-twin-based supply chain control. Int J Prod Res 59(19):5721–5742
54. Parrott A, Umbenhauer B, Warshaw L (2020) Digital twin: bridging the physical and digital. Deloitte Insights. https://www2.deloitte.com/xe/en/insights/focus/tech-trends/2020/digital-twin-applications-bridging-the-physical-and-digital.html. Accessed 8 Sept 2022
55. Pehlken A, Baumann S (2020) Urban mining: applying digital twins for sustainable product cascade use. In: Proceedings of the 2020 IEEE international conference, Cardiff, UK, 15–17 June 2020, pp 1–7
56. Petrova-Antonova D, Ilieva S (2019, June) Methodological framework for digital transition and performance assessment of smart cities. In: 2019 4th international conference on smart and sustainable technologies (SpliTech). IEEE, pp 1–6
57. Phillips A (2020) Is your business ready for a digital twin? TIBCO Blog. https://www.tibco.com/blog/2020/10/21/is-your-business-ready-for-a-digital-twin/. Accessed 8 Sept 2022
58. Piromalis D, Kantaros A (2022) Digital twins in the automotive industry: the road toward physical-digital convergence. Appl Syst Innov 5(4):65
59. Plautz J (2022) Las Vegas unveils digital twin of downtown. Constructiondive. https://www.constructiondive.com/news/las-vegas-unveils-digital-twin-of-downtown/617127/. Accessed 9 Sept 2022
60. Popa EO, van Hilten M, Oosterkamp E, Bogaardt MJ (2021) The use of digital twins in healthcare: socio-ethical benefits and socio-ethical risks. Life Sci, Soc Policy 17(1):1–25
61. Pylianidis C, Osinga S, Athanasiadis IN (2021) Introducing digital twins to agriculture. Comput Electron Agric 184:105942
62. Qi Q, Tao F (2018) Digital twin and big data towards smart manufacturing and industry 4.0: 360 degree comparison. IEEE Access 6:3585–3593
63. Quintanilla G (2022) What is digital supply chain twin and how can it support your strategic decisions? https://www.aimms.com/story/what-is-a-digital-supply-chain-twin-and-how-can-it-support-your-strategic-decisions/. Accessed 2 Sept 2022
64. Salman FS, Yücel E (2015) Emergency facility location under random network damage: insights from the Istanbul case. Comput Oper Res 62:266–281. https://doi.org/10.1016/j.cor.2014.07.015
65. Sanderse B, Weippl E (2018) Digital twins. ERCIM News 114:1–56. https://doi.org/10.1177/2399808318796416
66. Schleich BN, Anwer N, Mathieu L, Wartzack S (2017) Shaping the digital twin for design and production engineering. CIRP Ann Manuf Technol 66(1):141–144
67. Semenov Y, Semenova O, Kuvataev I (2020) Solutions for digitalization of the coal industry implemented in UC Kuzbassrazrezugol. In: E3S web of conferences, vol 174. EDP Sciences, p 01042
68. Siemens Healthineers (2019) The value of digital twin technology. https://marketing.webassets.siemens-healthineers.com/1800000007115634/143ec96042c1/Siemens-Healthineers_Whitepaper_Digital-Twin-Technology3_1800000007115634.pdf. Accessed 7 Sept 2022
69. Sivalingam K, Sepulveda M, Spring M, Davies P (2018, March) A review and methodology development for remaining useful life prediction of offshore fixed and floating wind turbine power converter with digital twin technology perspective. In: 2018 2nd international conference on green energy and applications (ICGEA). IEEE, pp 197–204
70. Srai JS, Settanni E, Tsolakis N, Aulakh PK (2019) Supply chain digital twins: opportunities and challenges beyond the hype
71. Tao F, Zhang H, Liu A, Nee AY (2019) Digital twin in industry: state-of-the-art. IEEE Trans Industr Inf 15(4):2405–2415. https://doi.org/10.1109/TII.2018.2873186.ISSN15513203

72. Tao F, Zhang M, Cheng J, Qi Q (2017) Digital twin workshop: a new paradigm for future workshop. Comput Integr Manuf Syst 23(1):1–9
73. Tzachor A, Richards CE, Jeen S (2022) Transforming agri-food production systems and supply chains with digital twins. arXiv preprint arXiv:2202.07455
74. Uhlemann TH, Schock C, Lehmann C, Freiberger S, Steinhilper R (2017) The digital twin: demonstrating the potential of real time data acquisition in production systems. Procedia Manuf 9:113–120. https://doi.org/10.1016/j.promfg.2017.04.043
75. Van der Valk H, Haße H, Möller F, Arbter M, Henning JL, Otto B (2020, August) A taxonomy of digital twins. In: AMCIS
76. van Houten H (2018) The rise of the digital twin: how healthcare can benefit. Philips News Center. https://www.philips.com/a-w/about/news/archive/blogs/innovation-matters/20180830-the-rise-of-the-digital-twin-how-healthcare-can-benefit.html. Accessed 7 Sept 2022
77. Wang Y, Wang X, Liu A (2020) Digital twin-driven supply chain planning. Procedia CIRP 93:198–203
78. Wong EY, Mo DY, So S (2021) Closed-loop digital twin system for air cargo load planning operations. Int J Comput Integr Manuf 34(7–8):801–813
79. Yitmen I, Alizadehsalehi S, Akıner İ, Akıner ME (2021) An adapted model of cognitive digital twins for building lifecycle management. Appl Sci 11(9):4276
80. Zhao G, Cao X, Xiao W, Zhu Y, Cheng K (2019, July) Digital twin for NC machining using complete process information expressed by STEP-NC standard. In: Proceedings of the 2019 4th international conference on automation, control and robotics engineering, pp 1–6

Industry 4.0 Technologies: Opportunities in the Sustainable Supply Chain Management

Deepak Datta Nirmal and K. Nageswara Reddy

Abstract The Fourth industrial revolution, i.e., Industry 4.0, is significantly transforming the business world in almost every sector. Digitalization of the supply chain, dynamic business environment, and changing customer needs have intensified the competition. At the same time, it is necessary to focus on the issue of rapid climate change and pollution due to manmade activities. Thus, the incorporation of sustainable practices for excelling in economic, social, and environmental dimensions of sustainability has become the key responsibility as well as the strategy of top-performing firms. Recently, research in analyzing the linkages between Industry 4.0 and sustainable supply chain management has gained momentum. The question of how Industry 4.0 technologies can aid in sustainable development has attracted the interest of researchers. Hence, it is essential to investigate how to achieve sustainable performance by using advanced technologies. The current chapter gives detailed information about the applications and opportunities offered by prominent Industry 4.0 technologies like "big data", "Internet of things (IoT)", "additive manufacturing", "blockchain technology," etc., in the field of sustainable supply chain management. The chapter provides a critical discussion, SWOT analysis, limitations of these technologies, and their huge potential to achieve sustainable performance.

Keywords Industry 4.0 · Supply chain · Sustainable practices · Blockchain technology · SWOT analysis

1 Introduction

The advent of Industry 4.0 technologies has opened the doors of opportunities in the dynamic business environment. This fourth industrial revolution and Industry 4.0 technologies have influenced every part of the supply chain. It is evident from the past literature that the use of technologies like "big data", "additive manufacturing",

D. D. Nirmal · K. N. Reddy (✉)
Operations and Supply Chain, Indian Institute of Management Jammu, Jammu & Kashmir, India
e-mail: knsr481@gmail.com

K E K Vimal et al. (eds.), *Industry 4.0 Technologies: Sustainable Manufacturing Supply Chains*, Environmental Footprints and Eco-design of Products and Processes,
https://doi.org/10.1007/978-981-99-4819-2_9

"blockchain", "Internet of things", "cloud computing", etc., have a positive effect on the competitive advantage and performance of organizations. Industry 4.0 technologies hold the promise to provide innovation, overall cost reduction, vertical as well as horizontal integration, real-time visibility, health, and safety to employees, and enhanced economic, social, and environmental performance. On the other hand, a rise in pressures from stakeholders, strict environmental laws and regulations, and an increase in consumer environmental awareness are some of the prime factors that drive industries to implement sustainable practices. Recently, it has been found that industries are trying to minimize the amount of energy and natural resource consumption for production. They are trying to produce products that have a minimum negative impact on the environment. Hence, knowing the implications of the Industry 4.0 tools in the sustainable supply chain area is crucial.

This chapter intends to give a detailed picture of Industry 4.0 opportunities, particularly in the sustainable supply chain (SSC) management area. The application of these technologies, individually or together, has varied impacts on different parts of the SSC. In addition, this chapter tries to give insights into how these technologies will help to achieve cleaner production, high quality, remanufacturing, recycling, innovation, flexibility, transparency, tracking, etc. The methodology adopted to understand the promising opportunities that lie in Industry 4.0 adoption for the SSC area is a systematic review and thematic analysis of the past literature. In addition, the recent reports, book chapters, and other literature are also considered for comprehensive analysis and insights. It is found that there is ambiguity among the researchers regarding the positive implications of Industry 4.0. However, many other academicians, industrialists, and researchers have proven that if used strategically, Industry 4.0 technologies have the potential to change the dynamics of the business world in the future. Industry 4.0 is no longer an option but is necessary to achieve a competitive advantage. Thus, analyzing the Industry 4.0 opportunities in SSC management is essential.

2 Industry 4.0

The world has been through different industrial revolutions comprised of mechanization, electrification, and computerization [1]. These revolutions have been changing industries and their operations throughout the world. The most recently discussed topic among Industry practitioners, researchers, and academicians is Industry 4.0. The term Industry 4.0 first appeared at the Hannover Fair in 2011. It originated through a German government initiative to advance the computerization of manufacturing [2]. Although many technologies are associated with Industry 4.0, some of the highly discussed and crucial technologies are depicted in the inner circle of Fig. 1. Industry 4.0 not only talks about the different technologies but also helps in synergizing these technologies.

It is visible that Industry 4.0 helps to achieve various things like increased flexibility, higher quality, high responsiveness, overall cost reduction, less pollution, fewer

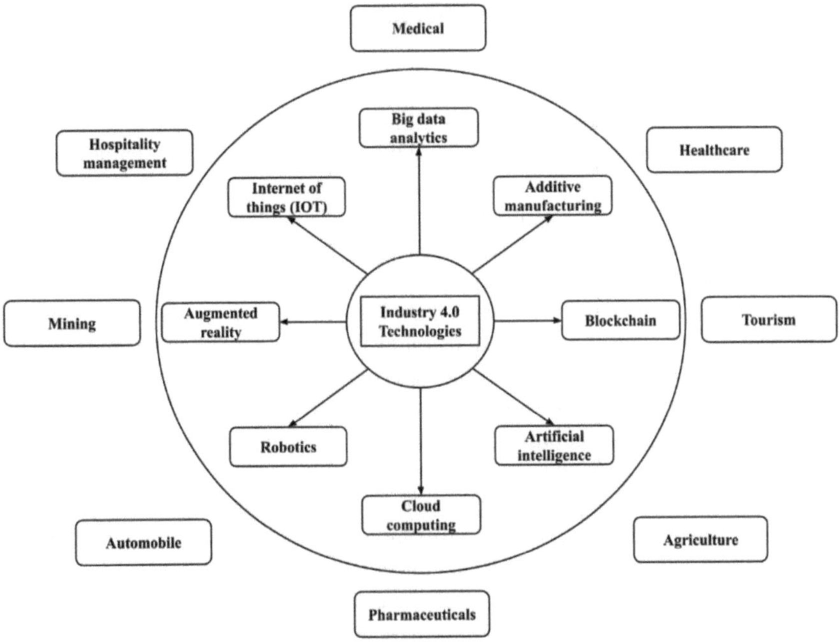

Fig. 1 System model of the proposed scheme

wastages, higher safety, innovation, etc. [3, 4]. Industry 4.0 is commonly accepted as an umbrella term for different technologies; however, its success will depend on the other dimensions, as discussed further. Organizations need to understand Industry 4.0 as the combination of "Technologies", "people and culture", and "strategy" [5, 6]. The "Technology" dimension includes various technologies, as given in Fig. 1. "People and culture" include the skill of the workforce and organizations' continuous improvement culture. The "Strategy" dimension comprises the ability of the organization to respond quickly, invest in Industry 4.0 technologies, make the plan for smart production, etc. It is also argued that organizations should develop their Industry 4.0 capabilities consisting of managerial, operational, and technological capabilities to gain higher organizational performance [4]. Thus, it is important to understand that successful implementation of Industry 4.0 technologies can be done only when the organizations focus on its overall dimensions like the workforce's skill, management attitude, investment, plans, etc., and develop their dynamic capabilities. As shown in the outer layer of Fig. 1, there are ample applications of Industry 4.0 that can be found in various domains despite the manufacturing sector. However, the present chapter will primarily concentrate on the opportunities and applications of Industry 4.0 technologies in sustainable manufacturing supply chains.

3 Sustainable Supply Chain Management

Sustainable supply chain (SSC) management is defined as "the management of material, information, and capital flows as well as cooperation among companies along the supply chain while taking goals from all three dimensions of sustainable development, i.e., economic, environmental and social, into account which are derived from customer and stakeholder requirements" [7]. Fundamentally, SSC management is about including different practices and strategies related to environmental, social, and economic aspects in conventional supply chain practices. It aids in achieving sustainable performance and deals with the stakeholders' pressures related to the environment. SSC is also considered an extension of the green supply chain. Initially, the incorporation of sustainability issues into industry practices was mainly driven by the various pressures from the government, competitors, customers, social groups, etc. However, including sustainability dimensions (environmental, social, and economic) in industry operations have been viewed as the strategy and source to attain competitive advantage. Green supply chain practices consist of green packaging, design, innovation, waste reduction, optimal utilization of resources, eco-efficiency, etc. At the same time, social sustainability is concerned with the employees' health and safety, well-being, equal rights to all, no discrimination based on gender, etc. For long-term sustainable performance and competitive advantage, industries should develop their capabilities and use their resources to implement sustainable practices. In addition, it is found that circular economic practices like recycling, remanufacturing, reusing, reducing, etc., have been beneficial to achieving sustainable organizational performance.

4 Industry 4.0 and Sustainable Supply Chain Management

Industry 4.0 and SSC management practices are the two revolutionary waves in the field of business management. Understanding the applications of Industry 4.0 technologies is essential to successfully implement sustainable practices in the industries [8]. Thus, recent studies have tried to analyze the various linkages between Industry 4.0 technologies and green, social, and other sustainable practices.

4.1 Opportunities and Applications of Prominent Industry 4.0 Technologies in SSC Management

4.1.1 Big Data Analytics and Sustainability

Big data are larger datasets with significant volume, huge diversity, and high velocity and are thus difficult to handle for traditional data processing systems. Due to the emergence of Industry 4.0 technologies, machine-to-machine interactions, and the use of robots, sensors, and automated machines, a large amount of data are being generated. Such large data are difficult to handle and analyze for the decision-makers. However, such big data have huge potential and high hidden values that industries can use in the future. Hence, various techniques like cloud computing, statistical analysis, and data analytics can be used to unlock the potential of big data. This confirms the importance of big data as an important asset for organizations. If organizations utilize this resource wisely, they can obtain a sustainable competitive advantage [9].

Companies such as Google, Amazon, and Flipkart have developed efficient big data analytics capabilities to gain an advantage and capture the potential of such high-speed large data. Organizations that have developed big data analytics capabilities, like good management skills for handling, storing, and managing the data, high technical skills, data-driven decision-making, learning culture, adequate resources, etc., have a high chance of achieving sustainable performance [10]. Despite the benefits of big data, there are many challenges to successfully implementing big data management, like data privacy, data security, legal requirements, data storage, etc. Thus, corporate commitment and support are essential for developing big data management capabilities. It is also found that employees with greater ability to use big data and higher technical skills produce organizational learning and sustainable performance [11]. Big data predictive analytics can help to forecast the development of green products and to effective design of such products. Decisions based on the data rather than the intuitions of the individuals will help industries with accurate procurement and supplies of inventories. Customers' demand can be traced and thus forecasted with more accuracy by using the past patterns of the data and effective data analysis. Big data analytics also helps reduce the time for regular operations of the company and enhances the organization's agility. The collaboration with the suppliers can also be strengthened by big data analytics due to real-time data processing and information sharing. Thereby, the development of big data analytics capabilities is crucial to gaining higher performance and competitive advantage.

In short, big data analytics offers the following opportunities for industries:

- Improved Health and safety practices can be obtained by using the predictive analytics capabilities of big data.
- Errors due to human intuitions in decision-making can be avoided by using data-based decision-making.
- Better supply chain integration and high organizational agility can be achieved.
- Better communication and information sharing.
- Reduction in costs, wastages, and environmental pollution occurring through business operations is possible using big data analytics.

4.1.2 Internet of Things (IoT) and Sustainability

IoT can be defined as a "technology which is intuitive, robust and scalable that enables the digital transformation of the connected world through the internet and communicates all the relevant information in real-time across the value chain" [12].

Thus, IoT is the technology that helps to connect physical objects and to exchange data and information with other objects/machines with the aid of the internet, sensors, software, and information system. The role and impact of IoT-enabled operations in organizations are crucial.

IoT has varied applications in different sectors. To name a few, IoT can be used to track the health of patients by using smart health wearable devices, to improve the productivity of the crops by monitoring them and assessing them related to the need for insecticides or fertilizers, and worker's safety can be enhanced by using alert messages from the machines and by predictive maintenance. IoT is beneficial to implement sustainable supply chain practices as it can help in tracking the products, which helps to perform complete retraction of the used products, and further for recycling, reuse, remanufacturing, etc. The interconnectedness of the machines provides high responsiveness to customer demands and high productivity. Recently, various IoT-based products have been commercialized due to their wide applications [12]. However, the major challenge of the IoT ecosystem is that it is primarily based on interconnected devices and systems that can communicate with each other over the Internet or other networks. For example, edge devices are essential IoT components in the sustainability context. For instance, we can install an IoT Edge device to monitor and control emissions using a Decentralized AI system that operates on these small devices. It means that all the devices in the IoT ecosystem need to go through the cloud. It required high storage capacity and processing power. In the future, this system may lack if a huge number of devices/machines get connected [13].

The summary of the benefits that can be obtained using IoT is as follows:

- Better visibility of the products can be achieved by using sensors throughout the supply chain.
- Smart production and supply chain automation can be developed by using various technologies like Radio Frequency Identification (RFID), wireless communication, cyber-physical systems, and Internet-connected devices.
- IoT provides real-time and accurate information sharing about the condition of the product or inventory, leading to the reduction in defects and improved performance.
- Better monitoring of the logistics operations can be obtained.
- Development of smart cars, smart cities, and effective monitoring of the vehicles, traffic, etc., can be achieved.

4.1.3 Blockchain Technology (BT) and Sustainability

Blockchain Technology (BT) is defined as a "distributed, shared, encrypted database that serves as an irreversible and incorruptible repository of information" [14, 15]. BT is a system designed; tamper-proof and prominent technology, which adds value to the supply chain network in product tracking, transparency in transactions, reduction in costs, and expediting the processes. Cryptocurrency, the crucial application of BT, has found tremendous opportunities in various sectors like tourism, real estate, supply chain transactions, and education. It is evident that companies like Toyota, Alibaba, De Beers, Walmart, Unilever, DHL, etc., use BT to manage and track their products and supply chain processes. It is perceived that BT can improve the performance of organizations by enhancing supply chain integration, process integration, and supplier integration [14, 16]. The key features of BT, like decentralization, immutability, and traceability, have been found to be of great use for enhancing the overall performance of the firms. The sustainability-related regulations, governments' rules, and policies, environmental laws, etc., can be effectively monitored by using the BT. Many stakeholders or suppliers fail to implement sustainable practices and reduce environmental emissions. However, they deceive the parent organizations by showing false claims. It may also happen in another way where reputed firms deceive the regulatory bodies by presenting false data. In such cases, the use of BT by providing the visibility of the processes to the concerned authorities can curb the failures and build trust in the system. The same can be applied in the case of monitoring labor-related issues, transactions across the supply chain network, and record keeping of the history of processes. [17]. The requirement of high computational power, lack of trust among the people about BT, and rigidity of the governments of many countries to allow BT-based transactions are some of the challenges to successfully adopt BT. Although implementation of BT is in the initial stage, primarily in the developing countries, its positive implications may be seen in the future.

In addition to the discussed benefits provided by BT, some of the crucial opportunities are listed below:

- BT can provide trusted transactions and avoid the risk of fraud between the parties.
- Environmental regulations and laws can be monitored by using BT.
- BT can be used for documentation of energy certificates and energy allowances and safeguard supply chain transactions.
- BT offers a variety of applications in healthcare, pharmaceutical, real estate, insurance, and other sectors due to features like transparency, traceability, immutability, etc.
- Companies that show falsified claims of performing and obeying according to environmental norms or social norms can be caught by using BT.

4.1.4 Additive Manufacturing (AM) and Sustainability

"AM, also known as three-dimensional (3D) printing, is a process that takes a digital 3D representation and produces the associated physical object layer by very thin layer, joining the layers as it goes along" [1]. AM is a prominent technology that has huge potential and applications in the mass customization of products, in the production of complex geometrical shaped products, intricate lattice structures, medical components like surgical guides, anatomical models, etc. Considering the priorities of operations management like quality, flexibility, speed, and cost, AM plays a better role than traditional manufacturing. There are many products that cannot be made through traditional manufacturing but can be made with ease by using AM.

Challenges in front of traditional production, for example, complex geometrical shaped products, require a lot of effort from the designing side to the formation of their prototypes, product-centric development of the specific fixtures, tools, etc., can be avoided by using AM. Thus, AM finds a lot of opportunities in various sectors, especially in the aerospace, medical, and production of complex parts in the automotive sectors. The wastage of the material can be potentially reduced by using the AM, which is opposite to traditional subtractive manufacturing, where the material is cut from the solid block to produce the products [18]. Despite the advantages of the AM, many challenges and disadvantages exist. Industry practitioners are facing the barriers like lack of knowledge regarding AM technology for consistent and stable production, expensive materials, complexities for converting metal alloys into the material (e.g., powder form) suitable for AM, and lack of proper services from the equipment vendor [19].

Following is the list of prominent benefits provided by the AM:

- AM can help to produce fixtures, tools, complex geometrical shaped products, and functional prototypes in a quick and cost-effective manner.
- Fewer material wastages, low inventory, and high flexibility can be obtained.
- Sustainability can be achieved due to the use of less amount of material for production as compared to the traditional manufacturing.

4.1.5 Artificial Intelligence (AI) and Sustainability

Artificial Intelligence (AI) is usually understood as a system in which machines/computers imitate human cognitive capabilities or intelligence for performing various human-like tasks like decision-making, face recognition, voice recognition, etc. AI helps to avoid repetitive tasks performed by a human that lead to mental stress. AI does it by using automation, integrating thousands of machines to solve intricate problems, and providing insights from the large data that is generated from the sensors, machine-to-machine interactions, emails, videos, etc. AI found its applications in many sectors like Agriculture, Automotive, Healthcare, Hospitality, and Tourism. AI enables higher sustainable performance through its various applications in energy distribution, sustainable transportation, environmental governance, smart cities, and smart factories. Sustainability, with all its dimensions, including social, economic, and environmental, is a complex issue. Thus, researchers need to understand stakeholders' behavioral responses and technical capabilities to effectively implement AI for sustainability. AI can be used in the areas where humans face difficulties in performing their work. These areas include mines, minerals, radioactive material extraction sites, tunnels, etc. [1, 20]. Despite AI's potential to enhance high sustainable performance, it can also lead to the generation of more carbon emissions and high energy consumption due to AI model training, dependencies on big data, etc. Thus, it is necessary to use it optimally with other technologies so that it aids in sustainable development.

Some of the benefits provided by using AI in the business world can be given as follows:

- Improvement in the quality of the job by avoiding the repetitive tasks of the employees.
- Prediction of dangerous incidents and risks by using AI algorithms is crucial for the effective implementation of social supply chain practices and employee welfare (like ChatGPT that designed to generate human-like text and carry out conversations with users).
- Error-free measurement of the quality of the products becomes possible by integrating AI for testing the quality of products.
- It can be used for efficient energy distribution, water resource conservation, and sustainable transportation.

4.2 SWOT Analysis

Figure 2 shows the SWOT analysis of the Industry 4.0-enabled SSC management. The points covered under the SWOT are incurred from the detailed discussion in the chapter.

Strengths (S): The content covered under this heading describes the success factors organizations must possess and develop to adopt SSC practices and emerging technologies.

Weaknesses (W): This area needs to be given attention to overcome the lacking. For example, employees need to be given proper training for the upgradation of their skills.

Opportunities (O): These are the advantages and benefits of applying Industry 4.0 tools and SSC practices. Organizations need to extend their abilities to avail these opportunities.

Threats (T): As both the Industry 4.0 and SSC management areas are new, there are a lot of uncertainties and ambiguities about their outcomes and proper way of application. Although these technologies help to gain sustainable performance, they should be optimally used because their application also consumes energy.

5 Summary

The chapter has shown the applications of Industry 4.0 technologies and enlisted the opportunities. Despite the huge potential of Industry 4.0 to enable SSC practices, they have certain limitations and disadvantages. There is no set of specific standard rules/processes for implementing Industry 4.0 technologies. Industries often lack expertise and knowledge about these technologies. Thus, limitations can be reduced once the knowledge about these technologies and their effective implementation is known to the industries. If the firms are committed to excelling in environmental sustainability, developing their dynamic capabilities, and investing in developing IT infrastructure, training, and development programs for employees, etc., then they

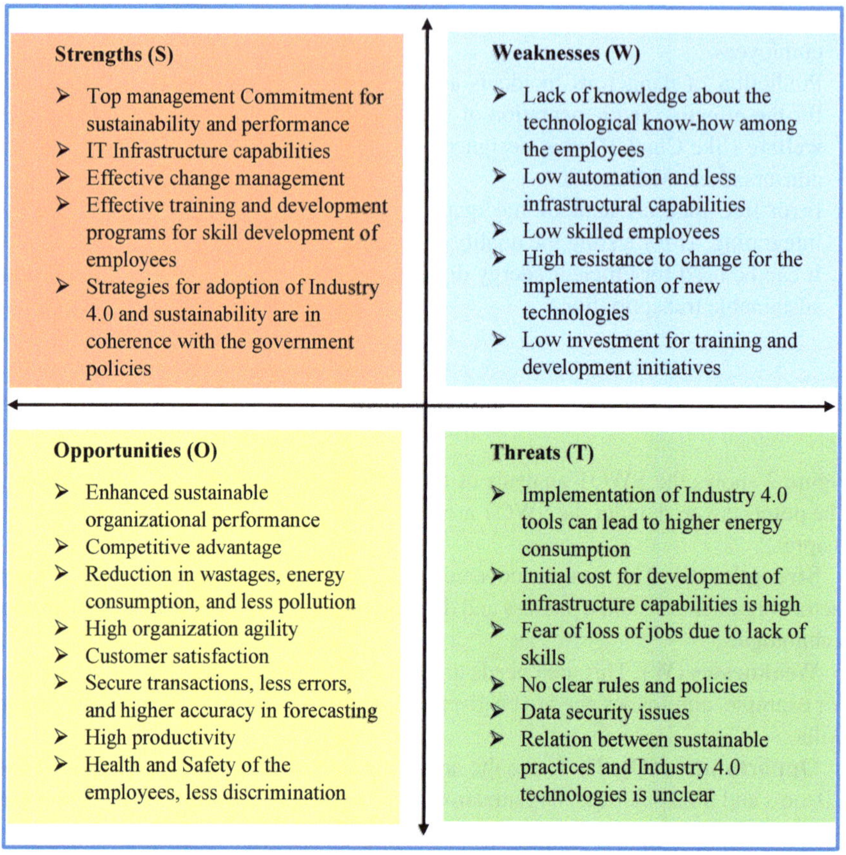

Fig. 2 SWOT analysis—Industry 4.0 enabled sustainable supply chain practices

can gain a competitive advantage over other firms. It is evident that Industry 4.0 technologies enable SSC practices in various ways as follows: reducing the wastage of perishable food products from regular tracking, monitoring the environmental and safety regulations, enhancing the responsiveness of the firms, and avoiding the risks of the employees for working in dangerous workplaces and reducing the pollutions.

In addition, it cannot be ignored that some researchers and academicians have raised some doubts regarding the need and consumption of very high energy due to the requirement of high storage capacity systems and high computation power machines for training datasets and other tasks. Thus, it needs to be clear that on one side, Industry 4.0-enabled SSC practices can lead to enhanced performance and competitive advantage, whereas, on the other side, it may initially lead to high energy utilization. At length, it can be concluded that Industry 4.0 technologies may cost the industries in starting phase in terms of energy and investment, but, in the long term, the huge potential of these technologies to attain sustainable performance is undeniable.

References

1. Olsen TL, Tomlin B (2020) Industry 4.0: opportunities and challenges for operations management. Manuf Serv Oper Manag 22(1):113–122. https://doi.org/10.1287/msom.2019.0796
2. Luthra S, Mangla SK (2018) Evaluating challenges to Industry 4.0 initiatives for supply chain sustainability in emerging economies. Process Saf Environ Prot 117:168–179. https://doi.org/10.1016/j.psep.2018.04.018
3. Edwin Cheng TC, Kamble SS, Belhadi A, Ndubisi NO, Hung Lai K, Kharat MG (2021) Linkages between big data analytics, circular economy, sustainable supply chain flexibility, and sustainable performance in manufacturing firms. Int J Prod Res, 1–15. https://doi.org/10.1080/00207543.2021.1906971
4. Belhadi A, Kamble S, Gunasekaran A, Mani V (2021) Analyzing the mediating role of organizational ambidexterity and digital business transformation on industry 4.0 capabilities and sustainable supply chain performance. Supply Chain Manag. https://doi.org/10.1108/SCM-04-2021-0152
5. Bibby L, Dehe B (2018) Defining and assessing industry 4.0 maturity levels–case of the defence sector. Prod Planning Control 29(12):1030–1043. https://doi.org/10.1080/09537287.2018.1503355
6. Erboz G, Yumurtacı Hüseyinoğlu IÖ, Szegedi Z (2022) The partial mediating role of supply chain integration between Industry 4.0 and supply chain performance. Supply Chain Manag 27(4):538–559. https://doi.org/10.1108/SCM-09-2020-0485
7. Seuring S, Müller M (2008) From a literature review to a conceptual framework for sustainable supply chain management. J Clean Prod 16(15):1699–1710. https://doi.org/10.1016/j.jclepro.2008.04.020
8. de Sousa Jabbour ABL, Jabbour CJC, Foropon C, Filho MG (2018) When titans meet – Can industry 4.0 revolutionise the environmentally-sustainable manufacturing wave? The role of critical success factors. Technol Forecasting Soc Change 132:18–25. https://doi.org/10.1016/j.techfore.2018.01.017
9. Singh SK, El-Kassar AN (2019) Role of big data analytics in developing sustainable capabilities. J Clean Prod 213:1264–1273. https://doi.org/10.1016/j.jclepro.2018.12.199
10. Jeble S, Dubey R, Childe SJ, Papadopoulos T, Roubaud D, Prakash A (2018) Impact of big data and predictive analytics capability on supply chain sustainability. Int J Logist Manag 29(2):513–538. https://doi.org/10.1108/IJLM-05-2017-0134
11. Bag S, Wood LC, Xu L, Dhamija P, Kayikci Y (2020) Big data analytics as an operational excellence approach to enhance sustainable supply chain performance. Resour Conserv Recycling 153. https://doi.org/10.1016/j.resconrec.2019.104559
12. Manavalan E, Jayakrishna K (2019) A review of Internet of Things (IoT) embedded sustainable supply chain for industry 4.0 requirements. Comput Ind Eng 127:925–953. https://doi.org/10.1016/j.cie.2018.11.030
13. Banafa A (2020) Blockchain technology blockchain technology and applications
14. Kamble SS, Gunasekaran A, Subramanian N, Ghadge A, Belhadi A, Venkatesh M (2021) Blockchain technology's impact on supply chain integration and sustainable supply chain performance: evidence from the automotive industry. Ann Oper Res. https://doi.org/10.1007/s10479-021-04129-6
15. White GRT (2017) Future applications of blockchain in business and management: a Delphi study. Strateg Chang 26(5):439–451. https://doi.org/10.1002/jsc.2144
16. Vyas A (2022) Sonali and Shukla, Vinod Kumar and Gupta, Shaurya and Prasad, Blockchain technology: exploring opportunities, challenges, and applications
17. Kshetri N (2021) Blockchain and sustainable supply chain management in developing countries. Int J Inf Manag 60. https://doi.org/10.1016/j.ijinfomgt.2021.102376
18. Spieske A, Birkel H (2021) Improving supply chain resilience through industry 4.0: a systematic literature review under the impressions of the COVID-19 pandemic. Comput Indus Eng 158. https://doi.org/10.1016/j.cie.2021.107452

19. Bromberger J, Ilg J, Miranda AM (2022) The mainstreaming of additive manufacturing. McKinsey & Company. https://www.mckinsey.com/business-functions/operations/our-insights/the-mainstreaming-of-additive-manufacturing
20. Nishant R, Kennedy M, Corbett J (2020) Artificial intelligence for sustainability: challenges, opportunities, and a research agenda. Int J Inf Manage 53:1–41. https://doi.org/10.1016/j.ijinfomgt.2020.102104

A Bibliometric Analysis of Smart Manufacturing and Way Forward

Saurabh Tiwari and Shantanu Trivedi

Abstract The term "smart manufacturing," often recognized as "intelligent manufacturing," remains widely utilized to indicate future manufacturing, or production of the future. It is an advanced style of manufacturing that blends industrial assets from the current and future perspectives with sensors, computer platforms, data-intensive modelling, communication technologies, management, simulation, and analytical engineering. It draws ideas from several fields, including data science, cloud computing (CC), artificial intelligence (AI), and cyber-physical systems (CPS). To give a comprehensive knowledge of the present understanding and many elements of smart manufacturing (SM), this study analyses the available literature, modern theories, information, and gaps for potential research initiatives. To determine the extent and trends of SM, a bibliometric study is utilized to reflect the various publishing sources, yearly publication numbers, keyword frequency, and top research and development regions.

Keywords Smart manufacturing (SM) · Circular economy (CE) · Industry 4.0 (I4.0) · Intelligent manufacturing · Sustainability

1 Introduction

Future production is often described by the phrase "smart manufacturing," often referred to as "intelligent manufacturing," which is used to describe such manufacturing [46, 56, 65]. In the area of smart manufacturing, publications are multiplying quickly. Numerous articles emphasize giving a detailed analysis of the problems

S. Tiwari (✉)
School of Business, University of Petroleum and Energy Studies, Dehradun, India
e-mail: tiwarisaurabht@gmail.com

S. Trivedi
Centre for Continuing Education, University of Petroleum and Energy Studies, Dehradun, India
e-mail: s.trivedi@ddn.upes.ac.in

K E K Vimal et al. (eds.), *Industry 4.0 Technologies: Sustainable Manufacturing Supply Chains*, Environmental Footprints and Eco-design of Products and Processes,
https://doi.org/10.1007/978-981-99-4819-2_10

affecting smart manufacturing. Several academics are interested in smart manufacturing, and they have published their findings in the literature. Smart manufacturing is a kind of production that uses optimized processes and procedures to increase yield while reducing energy footprint and costs. This is made feasible by the advancements in advanced modelling, controls, optimization, and big data that have occurred during the last decade. As a result, smart manufacturing is viewed as the industrial revolution 4.0.

Smart production systems are completely integrated, collaborative, and capable of real-time adaptation to changing plant circumstances and demands as well as supplier network and customer needs. Information and communication technologies are crucial to manufacturing systems. Cyber systems and associated intelligent and smart technologies [78, 87] still being in the development stage has led to the evolution and growth of CPS, Industry 4.0 (I4.0), digital twin, IoT, big data, cloud computing, and next-generation artificial intelligence [38, 54]. Several innovative manufacturing paradigms have been suggested to increase the "intelligence" or "smartness" of manufacturing systems and processes. "Smart manufacturing," a phrase that originated in the US but is now more widely used, is a collection of production processes that employ networked data, knowledge, and interaction tools to govern industrial activities. The application of AI techniques, such as big data analytics, business decision support systems, data-driven algorithms, and machine learning (ML), to improve industrial operations is referred to as "smart manufacturing." Although Smart Production primarily focuses on techniques for enhancing choices and processes inside industrial manufacturing settings, I4.0 primarily focuses on information sharing and interfaces.

The deliberation on the fundamental properties of CPS [66] gave an impression of the German I4.0 project and production activities in other nations [72]. The following scholars assessed the literature on smart manufacturing and the technology necessary for its development [35, 37, 41, 56]. The significance of data-driven industrial policymaking was emphasized by [31]. The investigations on the principles that impact smart manufacturing systems, products, and business aspects were done by [42].

Information and communication technology are crucial to manufacturing organizations. Future-proof AI, big data, I4.0, CPSs, IoT, cloud computing, [37, 75], digital twins (DT), CPSs are necessary to support the development of cyber systems and the related intellectual smart technologies [54]. Based on these ideas, a number of advanced manufacturing paradigms have been put out to increase manufacturing systems and processes as part of "intelligence" or "smartness" [46, 56, 86] or "Industrial Internet" (Bungart 2014), "Integrated Industry" [9], "Factory of the Future" [40], and "Smart Industry and Smart Manufacturing" [37]. This article seeks to advance the existing knowledge of Smart Manufacturing while presenting a novel angle for future research.

RQ 1. What is the present awareness and understanding of smart manufacturing?

RQ 2. What future research directions for smart manufacturing must be determined based on current works and recognized possible research gaps?

This research work is broken into eight categories. The first section gives a synopsis of the research area and describes the primary objective and directions we choose based on an examination of the literature depicted in the second section. The methods using the categorization of published literature from the Scopus database were drawn in the third section. Research analysis, conclusions, and findings were discussed in the fourth, fifth, and sixth sections, with respect to the area's significant topics, most illustrious journals, prolific writers, and scholarly and social structure. In the seventh section, a list of future research topics was provided, and in the eighth portion, this study's shortcomings were discussed.

2 Literature Review

Production companies are being propelled by I4.0 to transform into a modern production of CPSs that allow network-enabled smart manufacturing. The degree of "smartness" is mostly dependent on data-driven improvements that according to [37] put together all data about the manufacturing procedure available anytime, anywhere, and in a manner that is easy to realize throughout the organization and amongst linked companies. As SM turns into a fashion that influences industry and economic development, a lot of interacted equipment is applied more often to execute industrial tasks. Some of these devices, like a pipelined product line, heavily rely on the output from other equipment, while others could carry out the same or distinct functions or responsibilities. The link between networked equipment can also be energetically modified to improve tractability and adaptability to unique requirements. Therefore, the smart synergy of networked systems is crucial to increasing production system operation.

Thoben et al. [66] presented an outline of the I4.0 strategy of Germany as well as the manufacturing endeavours of other nations while addressing the fundamental properties of CPSs. In their analysis of the literature on smart manufacturing, [35] recognized technologies that are essential to its development. The conditions for data-driven industrial policymaking were outlined by [31]. The essential tools and impediments to the adoption of data-driven policymaking in business were recognized based on these demands. Standards that may have an impact on SM goods, procedures, and industry considerations were explored by [42]. O'Donovan et al. [53] focused on tasks confronting industrial data analytics functions. The writers proposed that formal methods should be used in place of prescriptive strategies to develop analytical abilities. According to [45], standards are essential for the incorporation of SM tools to address change concerns. It was recommended to use a mobile device-based technique to run enquiries in the provision of new updates. Zhang et al. [83] explored all the latest technologies such as high-speed computing, model-driven approaches, IoT, and cloud computing. A paradigm for industrial entity expertise representation that includes pertinent information and knowledge was proposed by [61]. The framework's role as a component of a CPS was demonstrated. The idea of smart manufacturing equipment that is controlled by wireless and Internet of Things technologies was put out by Zhong et al. [84]. Investigating the behaviour of SM products involved data analytics. Devices, computer programs, transmission tools, data-intensive modelling, management, virtual reality, and analytical engineering combine with the future industrial resources in smart manufacturing. Smart manufacturing makes use of CPSs, the IoT (and everything), cloud computing, model-driven computing, AI, and data science. When put into practise, these interrelated ideas and tools will make industry the defining feature of the following manufacturing transformation. Table 1 summarizes the definition of SM.

Table 1 Definition of smart manufacturing

Author(s)/ organization	Definition
[14]	The deployment of interacted value-based tools across the manufacturing and supply chain sector is known as smart manufacturing and is rapidly expanded and widespread. It simultaneously initiates and responds to a fundamental and drastic shift in the way business is conducted towards needs-based businesses, application based supply chain facilities, and widespread employee creativity and worker participation
[57]	SM is built on the concept of a smart plant and intends to enable effective, affordable, adaptable, and individualized mass manufacturing
[42]	Smart Production Processes (SMS) seek to optimize these capabilities via the deployment of modern tools that promote the quick movement and broad usage of digital information inside and across industrial structures. It is enabled by new stages of manufacturing responsiveness, excellence, and productivity across our facilities and businesses, enhancing long-term competitiveness
[37]	SM is about the independence, development, model, and optimization of the industrial organization, not the level of mechanization of the production floor. The level to which a manufacturing firm's basic business has been characterized in cyber space will define its degree of "smartness"
Mittal et al. [49]	SM was defined by five features: situation responsiveness, modularization, heterogeneousness, interactive, and architecture, as well as eleven technologies and three allowing features: legal and protocols
NIST (2017)	According to the definition of SM, these systems are "completely integrated, collaborative manufacturing systems that adapt in real time to changing demands and conditions in the plant, in the supplier network, and in consumer wants"
Abubakr et al. [1]	Utilizing the most recent advancements in AI, Cloud Computing and the IoT, the smart manufacturing concept (IoT)
Wang and Gao [79]	Computer-integrated manufacturing and AI are two examples of advanced manufacturing techniques that allow data-enabled flexibility through the manufacturing phase, from merchandise pattern through method development, management, and optimization to merchandise excellence policy

3 Methods

An objective method of studying patterns is done using Bibliometric analysis, related to a research field's included disciplines, keywords, authors, journals, institutions, and documents [4, 76]. Scopus database is used between 1996 and 2022 for the applications of thorough science mapping analysis. The bibliometric analysis is performed using bibliometrix package for the R programming language. The quantitative tool used for bibliometrics research and analysis is Bibliometrix software. The "Scopus" database's bibliographic information is imported using this software. This programme may also be used to build different kinds of network analysis. It is applied to present science mapping assessment by utilizing the Bibliometrix package capabilities and the Shiny user interface, Biblioshiny.

We examine the output and effects of search areas like "smart manufacturing" in research using performance analyses, and we search the literature for research topics using scientific maps. The search was started with a list of the fields that are involved in research on SM. The study area can be interdisciplinary or multidisciplinary, if it is associated with more than one discipline. The next step was to analyse the relevancy of the published venues using citation assessment of various journals. Co-citation evaluation of the journals uses the frequency of journal citations in other publications to pinpoint research topics. The citation analysis of the author was performed to examine the study output of the authors [12]. To identify themes in their works, the author used co-citation analysis. When multiple authors are conjointly referenced in another publication, this is known as a co-citation association. Co-citation analysis thus enables the classification of study issues, that are of particular relevance to quoting authors. Additionally, it enables the development of systems amongst leading scholars in a particular area (Rosetto et al. 2018). Based on the quantity of citations that a university's publications on smart manufacturing received, institution citation examination was utilized to observe each institution's research output. The connections between research institutes have been studied. The latent capabilities of smart manufacturing are the main subject of this study. Using keyword co-occurrence examination is an alternative way to spot research groups. This approach seeks to determine how frequently particular terms are used in conjunction. Using document citation analysis, journal manuscripts were tracked to ascertain their perceived usefulness. Researchers employed article co-citation examination to classify recurring subjects.

3.1 *Search String*

Data to be evaluated was gathered for the studies through the Scopus database regarding the population of all works on SM published between 1996 and 2022. The major objective is to classify publications that examine the development of smart manufacturing. The results guided the selection of the subsequent groups of

Table 2 Search string

Source	Search String
Data extracted on 5 December 2022 from Scopus database	'Smart AND manufacturing' OR 'smart AND production' OR 'smart AND factory' OR intelligent AND manufacturing OR industrial AND internet AND of AND things OR integrated AND industry OR factory AND of AND the AND future OR smart AND industry AND (LIMITTO (DOCTYPE, "ar")) AND (LIMITTO (SUBJAREA, "BUSI")) AND (LIMIT TO (LANGUAGE, "English")) AND (LIMIT-TO (SRCTYPE, "j"))

keywords, with the requirement that they appear somewhere in the search chain. The phrases "smart manufacturing," "smart production," "smart factory," "intelligent manufacturing," "industrial internet of things," "integrated industry," "factory of the future," and "smart industry" are most popular and well-known in searches, respectively. As mentioned previously, these keyword groups are searched and are mostly located in the titles or abstracts of the database publications we were looking for. We did a search on 5 December 2022, and discovered one list containing English-language materials from 1989. Table 2 provides more information on the search syntax utilized in the investigation.

The above-mentioned search string was used and it should be covered either in abstract, title, or keyword methods, is restricted to the English language, ans is found using the advanced exploration options. Simply copying and pasting the syntax on the Scopus database as needed will fetch the results. However, the likelihood of the conclusion being unchanged is essentially non-existent because the digital data is updated continuously.

3.2 *Database Selection and Collection of Data*

The Scopus database was chosen since it is a commonly used and acknowledged resource for scholars to undertake this sort of study. The above-mentioned keywords were searched for in Scopus titles and abstracts, yielding 1989 articles between 1996 and 2022.

4 Analysis

The section contains findings related to the study topic provided at the conclusion of the introduction.

4.1 Sources

In the explanatory assessment, the overall number of manuscripts, year-over-year, evolution trend, highly pertinent journals, h-index, and source progress are all provided (Table 3). According to the data, the search yielded 1989 papers from 2354 authors between 1996 and 2022.

The annual publication over the previous 26 years is displayed in Table 4 and Fig. 1. The growth in the number of publications is growing at a rate of 29.19 per cent every year. Over the past ten years, publications have been continuously rising in number. The pattern indicates that starting in 2014, there will be more than 20 publications per year, with a cap of 669 publications in 2021. This shows that the area is still developing and that, in the years to come, there will be an increase in publications.

The publishing pattern throughout time is seen in Fig. 1. The amount of research on smart manufacturing has clearly increased since 2014 (n26). The year 2021 saw a modest amount of paper published on smart manufacturing resulting in 669 articles.

Publications of the top journals on SM are shown in Fig. 2. The picture clearly shows that the top five journals with the most papers published are the JCP, TFSC, IJPR, IJPE, and PPC.

Table 5 below includes information on the most referenced journals in addition to the data shown in Fig. 2. The top five journals most frequently referenced in the area of SM are IJPR, IJPE, JCP, IJIM, and Procedia Cirp.

The journal with the highest h-index is shown in Fig. 3. IJPR, JCP, TFSC, IJPE, Production Planning and Control, and Industrial Management and Data Systems are six journals that have a h-index of more than 20.

To determine the most popular journals on this subject, an analysis was done. The most popular journals from 2016 are "Technological Forecasting and Social Change" and "Journal of Cleaner Production," according to Fig. 4. The top three journals with the most papers published are JCP (n = 147), TFSC (n = 144), and IJPR (n = 132).

Table 3 Summarized data

Main information	
Time frame	1996:2022
Total manuscript	1989
Keywords Plus (ID)	6374
Authors	5098
Per document average citation	43.04
Per document per year average citation	11.98
Document by Single-authored	155

Source Authors' expansion

Table 4 Yearly publications

Year	Number. of documents
1996	1
1999	1
2000	1
2001	1
2002	1
2003	2
2004	5
2005	1
2006	2
2007	6
2008	7
2009	9
2010	12
2011	3
2012	15
2013	16
2014	26
2015	38
2016	54
2017	78
2018	108
2019	196
2020	263
2021	669
2022	474

Source(s) Author's specific creation

4.2 Highly Prominent Authors and Keywords

Information about the most influential authors is provided in this section. The authors with the most documents in the field of SM are Zhang, Y., Kumar, A., Liu, Y., Dwivedi, Y. K., Javaid, M., Kumar, V., Gunasekaran, A., Liu, W., Haleem, A., Chen, Y., Huang, G. K., and Li, X, are the authors with the most publications in the field of SM, as shown in Fig. 5.

Additionally, the most frequently mentioned articles are shown in the Table 6. The findings show that the article by [24] in the TFSC, Xu et al. [81] in the IJPR, [60] in the IJPR, [23] in the IJPE, and others, published in the IJPR, [33] in Journal of Business Venturing, [35] in International Journal of Precision Engineering and Manufacturing-Green Technology, [29] in Electronics Markets, and [13] in IJPE

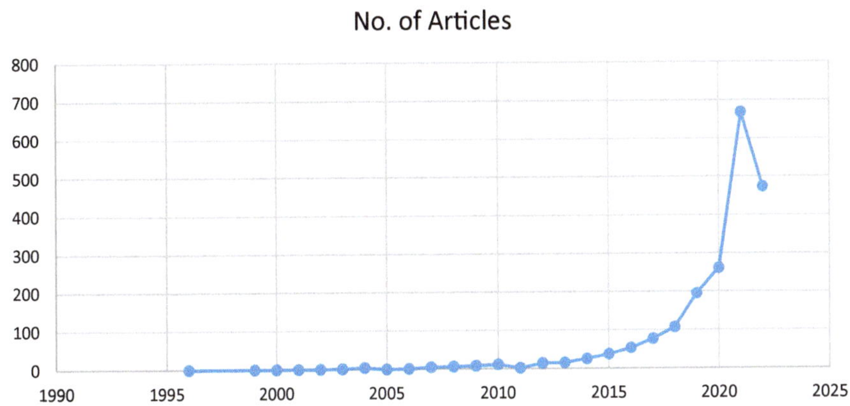

Fig. 1 Published document per year. *Source(s)* Author's specific creation

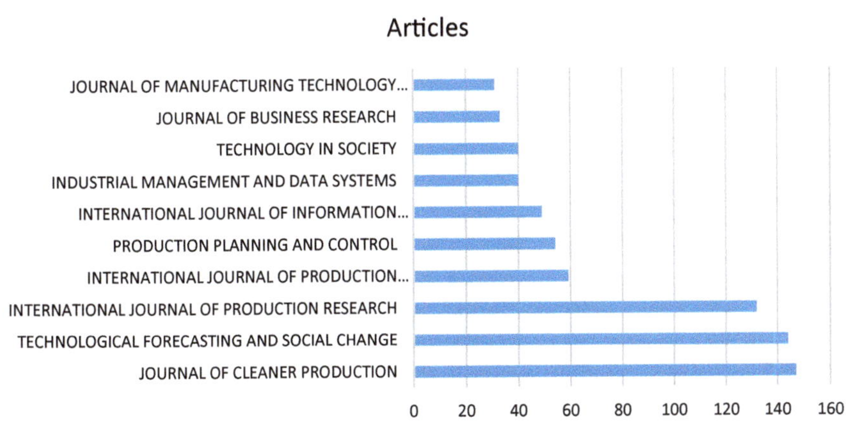

Fig. 2 Top Relevant Journals. *Source(s)* Author's own creation

are the seven most cited authors who have advanced the field with more than 700 document citations.

Co-word analysis is the most beneficial technique for comprehending the theoretical framework of the study conducted on a certain topic. The most often occurring words in the research paper are determined using a similar methodology. The most popular terms in the area are shown in Table 7. The outcome reveals that the keyword "Industry 4.0" is the most popular one.

The analysis of co-words reveals that "Industry 4.0," "Internet of Things," "Artificial Intelligence," "Big Data," "Blockchain," "Supply Chain Management," "Sustainability," and "Internet of Things" make maximum recurrently used keywords in the research papers (Fig. 6).

Table 5 Highest cited journals

Journals	Citation
International Journal of Production Research (IJPR)	3072
International Journal of Production Economics (IJPE)	2046
Journal of Cleaner Production (JCP)	1535
International Journal of Information Management (IJIM)	1234
Procedia Cirp	1230
Sustainability	1120
Journal of Business Research (JBR)	1103
IEEE Access	1034
Technological Forecasting and Social Change (TFSC)	1030
MIS Quarterly	983

Source(s) Author's own elaboration

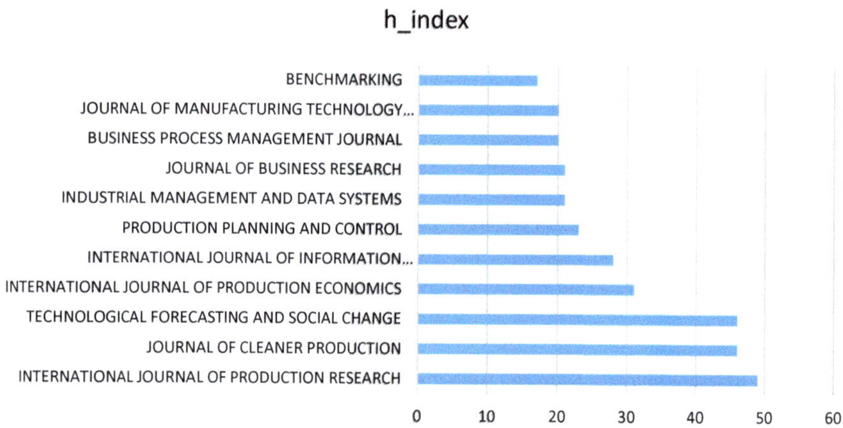

Fig. 3 High Impact Journals. *Source(s)* Author's specific creation

The most common terms used in the research were also revealed by analysing the current issues (see Fig. 7). For instance, the study reveals that the most often used phrases in the subject field are I4.0, decision-making, IoT, supply chain management, manufacturing, Big Data, Blockchain, and the AI.

A country-wise study (Fig. 8) was done to determine which nations contributed the most writers to papers about smart manufacturing. In this analysis, the names of nations with matching writers who have written influential works in this area were included. The nations that contributed the most to the article on smart manufacturing include China, India, the UK, the United States, Italy, Germany, Australia, Brazil, and Finland.

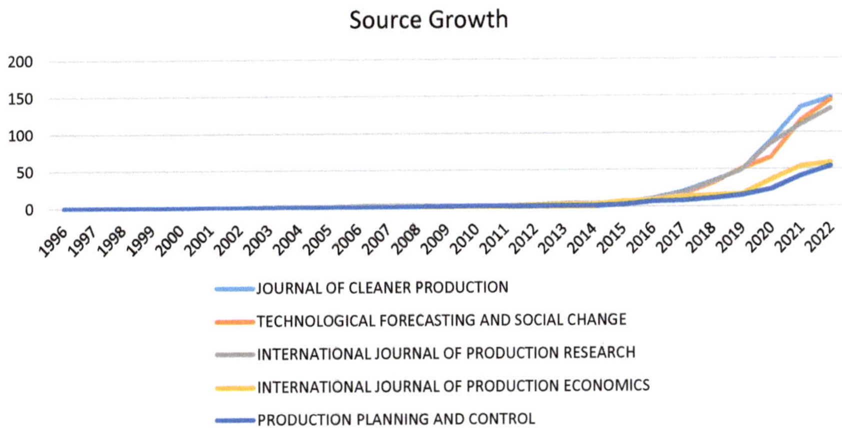

Fig. 4 Most trending Journals. *Source* Authors' elaboration using Biblioshiny

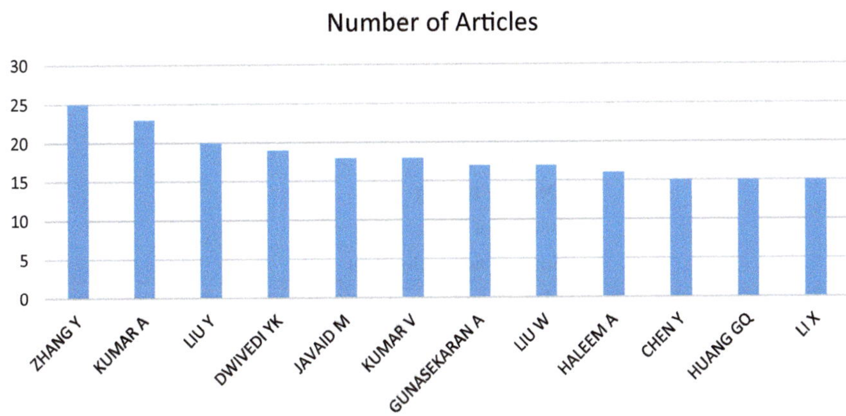

Fig. 5 Authors with a number of articles. *Source* Authors' elaboration using Biblioshiny

The study of citations for various nations is performed to determine which nations have the most citations in SM. The United Kingdom is in first place, with over 11,998 citations in the previous 26 years. China is in second place, with around 9821 citations. The United Kingdom, China, the United States, Italy, Germany, India, Brazil, and Korea are amongst the eight nations with more than 3000 citations. Brazil was discovered to be at the top of the average article citations study, having 85.5 citations. Table 8 displays the specifics of different nations' citations as well as the average article citations.

Table 6 Highest cited articles

Articles	Total citations	TC per Year
[24]	1838	306.333
Xu et al. (2018)	1333	266.6
[60]	1020	255
[23]	884	221
[33]	792	66
[35]	777	111
[29]	756	94.5
[13]	715	143
[80]	698	26.222
[30]	619	88.429
[32]	586	146.5
[37]	579	115.8
[51]	545	36.333
[27]	543	108.6

Source Authors' elaboration

Table 7 Most popular terms

Words	Occurrences
Industry 4.0	235
Decision-making	184
Internet of Things	165
Sustainable development	156
Manufacture	142
Supply chains	128
Supply chain management	123
Blockchain	102
Big data	101
Artificial intelligence	95
Design/methodology/approach	93
Technology adoption	82
Industrial research	80
Information management	75
Innovation	73
Technological development	72
Data analytics	64
Smart city	63
Competition	59
Embedded systems	57

Source(s) Authors' elaboration

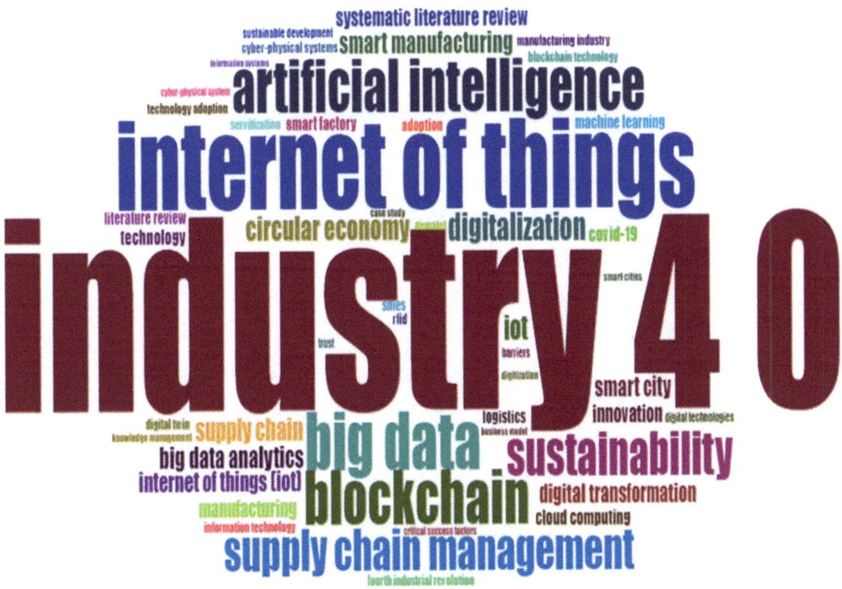

Fig. 6 Common words. *Source* Authors' elaboration using Biblioshiny

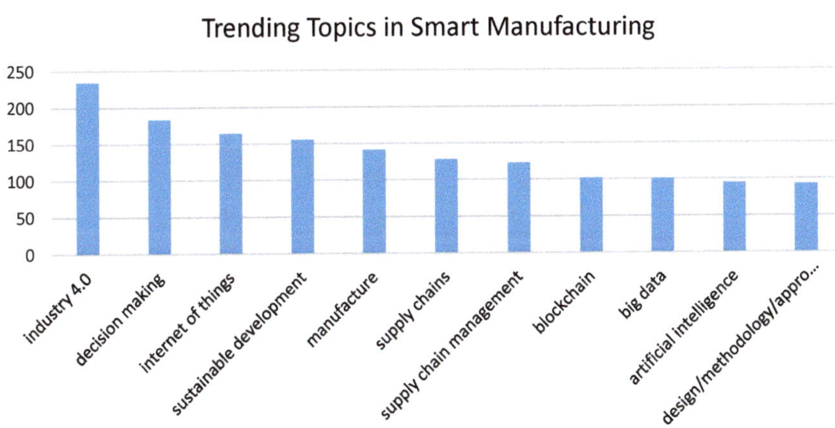

Fig. 7 Keyword productivity. *Source* Authors' elaboration using Biblioshiny

4.2.1 Conceptual Structure

The heat map visualization is a reliable method of determining the intensity of relationships between keywords. Since VOSviewer software provides a powerful GUI, a density map was constructed. Distinct colours in the SM term co-occurrence heat map (Fig. 9) represent distinct intensity standards. The more often used notion or topic is indicated by a higher density yellow colour. For example, "Industry 4.0" and "Internet

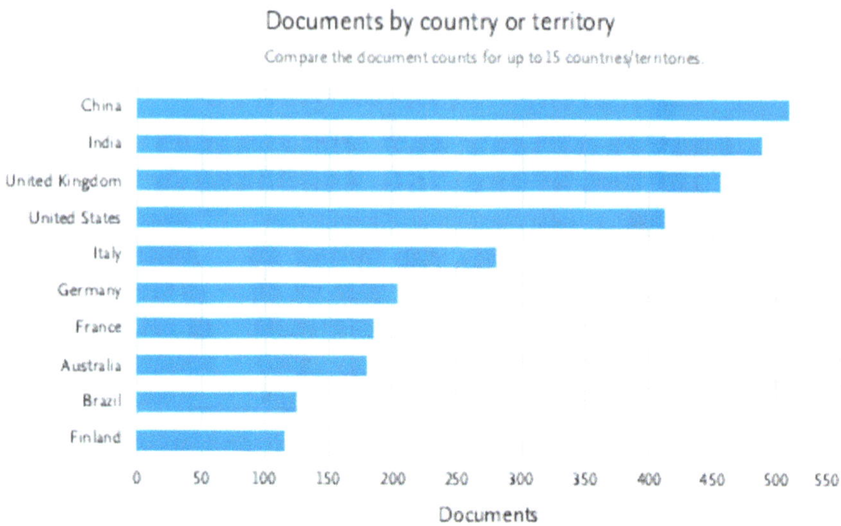

Fig. 8 Publications country-wise. *Source* Authors' elaboration

Table 8 Top-most cited countries

Country	Total citations	Average article citations
United Kingdom	11,998	69.8
China	9821	37.9
USA	8243	74.3
Italy	4897	37.4
Germany	3934	53.2
India	3927	25.5
Brazil	3678	85.5
Korea	3608	76.8
France	2610	49.2
Hong Kong	1707	53.3
Finland	1493	43.9
Sweden	1448	65.8
Malaysia	1421	52.6
Australia	1371	29.2
Iran	1286	44.3
Canada	1091	37.6

Source Authors' elaboration

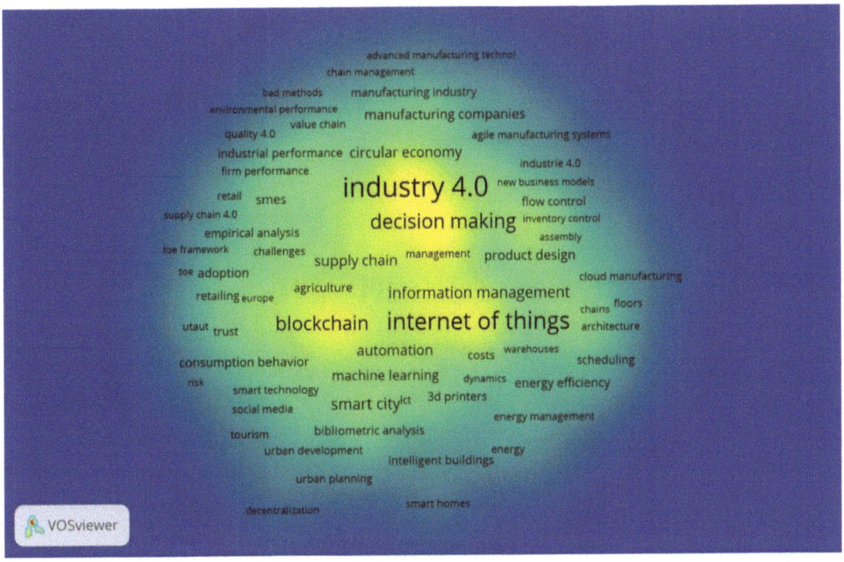

Fig. 9 Heat map of SM. *Source* Authors' elaboration using VOSviewer

of Things" have the maximum yellow colour density and are hence the most important terms. Aside from these two phrases, the greater intensity yellow colour can also be found on "Blockchain," "Decision-Making," "Information Management," "Supply Chain," and "Automation." The central subject in smart manufacturing research is the role of I4.0 in manufacturing over the use of the IoT, Blockchain, and information management.

Interpreting and understanding the arrangement and examining the subjects through keyword co-occurrence is another method. As Fig. 10 and Table 9 show, as well as additional assessment, five groups stand out. First is "Industry 4.0," which encompasses "Smart Manufacturing," "Technology Adoption," "Innovation," "Digitalization," "Global Value Chain," and "Additive Manufacturing." The next group includes "Artificial Intelligence," "Big Data," "Smart Factory," "Internet of Things," "Cyber physical system," "Big Data Analytics," and "Sustainability." 3rd group is "Competitive Advantage," "Digital Transformation," "Digital Technologies," "Digitization," "Manufacturing Industry," and "Servitization". In the 4th group "Emerging Economies," "Lean Manufacturing (LM)," "Lean Production (LP)," "Organizational Performance," and "Production Management" are considered as the main topic. The terms "Circular Economy (CE)," "Sustainable Development (SD)," and "Sustainable Manufacturing" make up the 5th and last group.

We may draw the conclusion that there are essentially two study streams that arise from the analysis of these five groups. The first stream is mostly technological and connects I4.0 with manufacturing. It uses big data, CPS, AI, and digital manufacturing transformation to achieve SM through enhanced organizational implementation and operation management. The 2nd is aimed at attaining manufacturing

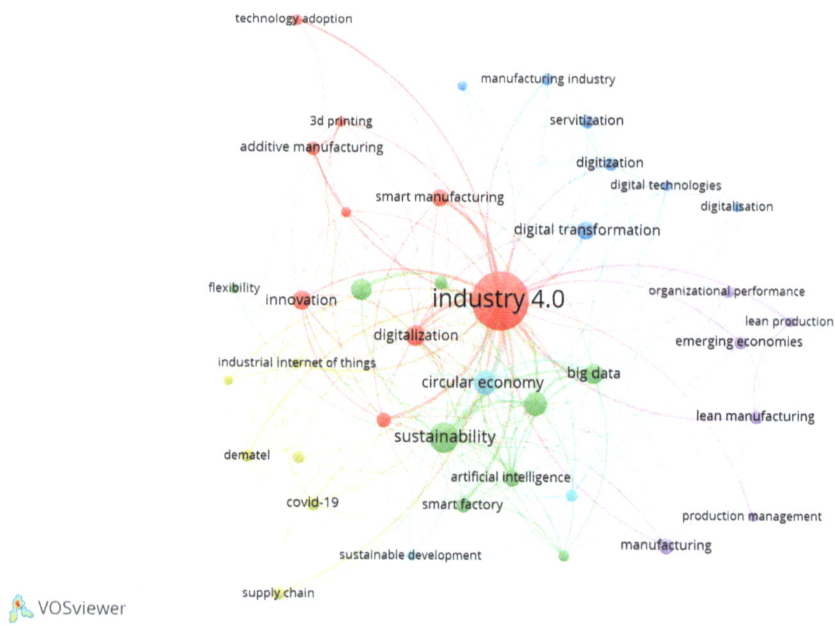

Fig. 10 The network of co-occurring keywords

Table 9 The group keywords

Group 1	Group 2	Group 3	Group 4	Group 5
I4.0	AI	Competitive advantage	Emerging economies	CE
SM	Big data	Digital transformation	LM	SD
Technology adoption	Smart factory	Digital technologies	LP	Sustainable manufacturing

Source Authors' elaboration using VOSviewer

sustainability through the adoption of CE, SD, and lean manufacturing, which are important for improving organizations' environmental implementation and providing competition to enterprises.

5 Findings

Around passing through an embryonic stage, the review on SM is expanding and attracting interest from both academia and business after 2015. The work significantly extends and adds to the form of understanding of SM (Buchi et al. 2020). The research adds to and improves the literature on SM by recognizing important writers, relevant themes, and the most significant publications in the field. The findings show that the most convincing research was directed by a select group of authors, including Frey, C. B., Xu, L. B., Saberi, S., Frank, A. J., Jones, K. V., Gretzel, U., and Dalenogare, L. S. Since 2015, there has been a multi-fold growth in academic interest in the topic of smart manufacturing, according to the trend of all 1939 papers. Most articles in the research field were written by the following authors (e.g., Zhang, Y; Kumar, A; Liu, Y; Dwivedi, Y. K; Javaid, M; Kumar, V; Gunasekaran, A., Liu, W; Haleem, A; Chen, Y; Huang, G. K; and Li, X). Consequences associated with pertinent authors, journals, citations, and associations in the area of Smart manufacturing reveal that the International Journal of Production Research, Journal of Cleaner Production, Technological Forecasting and Social Change, International Journal of Production Economics, Production Planning and Control, and Industrial Management and Data Systems arose as the greatest important journals in the area. An assessment of the associations and countries suggests that Hong Kong Polytechnic University, The University of Hong Kong, Indian Institute of Technology Delhi, University of Tehran National Institute of Industrial Engineering, and University of Johannesburg are the highest participating institutes. Additionally, the countries, which provided the maximum articles related to SM are the UK, China, the USA, Italy, Germany, India, Brazil, and Korea and have more than 3000 citations. According to group analysis, there are basically two research streams that make up the literature on smart manufacturing. The use of AI, big data, CPS, and digital transformation in production is part of the first stream, which is heavily focused on technology and connects I4.0 with manufacturing. SM is achieved due to improved execution at equally the organizational degree and fabrication management. The next focuses is on attaining sustainability in production beyond the adoption of lean manufacturing (LM), the circular economy, and sustainable development, all of which are important for improving companies' environmental performance and giving businesses a competitive edge.

6 Contributions and Implications

This study adds to the essence of understanding the subject of SM by compiling a list of the extremely influential authors, the highly pertinent and referenced journals, the highly quoted papers, promising keywords, and research groups. By emphasizing the key terms that make up the central part of the study for these topics and providing original and probable instructions for further investigation, the review also subscribes to the form of information on SM and sustainability.

Production is a source of the goods and facilities necessary for individual well-being, security, and comfort. From the perspective of both organizations and modern society, production is tied to all social events. Due to their role in producing goods that are crucial to both the quality of human existence and the health of the international financial system, industrial processes should be carefully examined in the framework of sustainability. As a consequence of the necessity for sustainable manufacturing practises in the present industrial revolution, the current research focuses on smart manufacturing. Additionally, a framework must be created for smart manufacturing for both practitioners and academics. The possible advantages of the SM method that the production industry is dealing with don't seem to be well understood by many businesses. The literature study shows that there has been a substantial surge in the importance of smart manufacturing since 2015, which is clearly obvious in the paper's focus on the topic. Certain strategies would be developed to try to raise the production area's proficiency. This approach helps to boost future business projections for the industrial sectors while also improving the condition of the ecosystem for potential productions. By offering thorough information on the researchers, articles, periodicals, and potential upcoming study issues, it also aids future research.

7 Future Research

The term "smart manufacturing" currently only refers to specific industrial companies and locations (maximum researches are in the United Kingdom, China, the USA, Italy, Germany, India, Brazil, and Korea). However, it is possible to expand it to other regions of the world. We recommend the following study areas based on our evaluation and subsequent analysis.

1. The adoption of smart manufacturing is driven by a big data system, thus industrial data must be properly gathered and processed. To create a more effective division of labour between intelligent robots and people, significant financial investments including advanced scientific equipment for vast data storing, recovery, handling, and assessment are required.
2. There isn't any clear explanation of "smart manufacturing" to increase manufacturing and sustainability understanding amongst producers, dealers, and consumers. There is a substantial difference amongst engineering and academic study in the subject of SM.
3. The innovative SM arrangement's complexity should be decreased to enhance interoperability with other environments. The social acceptability of factories 4.0 can be raised by providing training to employees on how to use factory 4.0 principles effectively. The strongest human resistance arises when workers are required to cooperate with robots and admit that computers can execute greater level cognitive functions.

8 Limitations

The current research has significant limitations, as do all others. To start, this assessment is thorough but not meticulous. The Scopus database is used in the study. We advise leveraging databases like Web of Science, EBSCO, and others for absolute and thorough analysis in future studies. Obtaining samples from many databases will greatly enhance the study. Increasing the relevance of the terms used to search the database will strengthen the search and enrich the manuscript. Researchers looking into smart manufacturing may find the study's findings useful regarding the investigation background and asperity. Next, we restricted our research to academic journal articles, eliminating theses, book chapters, and reports.

Other credible sources can be used to get further knowledge. Furthermore, while we made every effort to be trustworthy and inclusive, the subsequent evaluation may be theory-driven. Last, but not the least, these discoveries can serve as a springboard for future study into the domains of smart manufacturing.

References

1. Abubakr M, Abbas AT, Tomaz I, Soliman MS, Luqman M, Hegab H (2020). Sustainable and smart manufacturing: an integrated approach. Sustainability 12(6):2280
2. An XY, Wu QQ (2011) Co-word analysis of the trends in stem cells field based on subject heading weighting. Scientometrics 88(1):133–144
3. Bag S, Telukdarie A, Pretorius JHC, Gupta S (2021) Industry 4.0 and supply chain sustainability: framework and future research directions. Benchmarking Int J 28(5):1410–1450
4. Bahuguna PC, Srivastava R, Tiwari S (2022) Two-decade journey of green human resource management research: a bibliometric analysis. Benchmarking Int J. https://doi.org/10.1108/BIJ-10-2021-0619
5. Bai C, Dallasega P, Orzes G, Sarkis J (2020) Industry 4.0 technologies assessment: a sustainability perspective. Int J Prod Econo 229, 107776
6. Bocken NM, Short SW, Rana P, Evans S (2014) A literature and practice review to develop sustainable business model archetypes. J Clean Prod 65:42–56
7. Bressanelli G, Adrodegari F, Perona M, Saccani N (2018) Exploring how usage-focused business models enable circular economy through digital technologies. Sustainability 10(3):639
8. Büchi G, Cugno M, Castagnoli R (2020) Smart factory performance and industry 4.0. Technol Forecasting Soc Change 150:119790
9. Bürger T, Tragl K (2014) SPS-Automatisierung mit den Technologien der IT-Welt verbinden. Industrie 4.0 in Produktion, Automatisierung und Logistik: Anwendung· Technologien Migration 559–569
10. Chawla RN, Goyal P (2022) Emerging trends in digital transformation: a bibliometric analysis. Benchmarking Int J 29(4):1069–1112.
11. Cobo MJ, López-Herrera AG, Herrera-Viedma E, Herrera F (2011) Science mapping software tools: review, analysis, and cooperative study among tools. J Am Soc Inform Sci Technol 62(7):1382–1402
12. Culnan MJ (1986) The intellectual development of management information systems, 1972–1982: a co-citation analysis. Manage Sci 32(2):156–172
13. Dalenogare LS, Benitez GB, Ayala NF, Frank AG (2018) The expected contribution of Industry 4.0 technologies for industrial performance. Int J Prod Econ 204:383–394
14. Davis J, Edgar T, Graybill R, Korambath P, Schott B, Swink D, Wang J, Wetzel J (2015) Smart manufacturing. Annu Rev Chem Biomol Eng 6:141–160
15. Dubey R, Singh T, Ali SS, Tiwari S (2015) Contextual relationship among antecedents of truck freight using interpretive structural modelling and its validation using MICMAC analysis. Int J Logist Syst Manag 20(1):42–58
16. Dubey R, Singh T, Tiwari S (2012) Supply chain innovation is a key to superior firm performance an insight from indian cement manufacturing. Int J Innov Sci 4(4):217–230

17. Elkington J (1998) Partnerships from cannibals with forks: The triple bottom line of 21st-century business. Environ Qual Manage 8(1):37–51
18. Elkington J (2004) Enter the triple bottom line. triple bottom line: does it all add up 11(12):1–16
19. Erro-Garcés A (2021) Industry 4.0: defining the research agenda. Benchmarking Int J 28(5):1858–1882
20. Evans PC, Annunziata M (2012) Industrial internet: Pushing the boundaries. General Electric Reports, 488–508
21. Fatorachian H, Kazemi H (2018) A critical investigation of Industry 4.0 in manufacturing: theoretical operationalisation framework. Production Planning Control 29(8):633–644
22. Ferreira MP, Santos JC, de Almeida MIR, Reis NR (2014) Mergers & acquisitions research: a bibliometric study of top strategy and international business journals, 1980–2010. J Bus Res 67(12):2550–2558
23. Frank AG, Dalenogare LS, Ayala NF (2019) Industry 4.0 technologies: implementation patterns in manufacturing companies. Int J Prod Econ 210:15–26
24. Frey CB, Osborne MA (2017) The future of employment: How susceptible are jobs to computerisation? Technol Forecast Soc Chang 114:254–280
25. Garetti M, Taisch M (2012) Sustainable manufacturing: trends and research challenges. Prod Planning Control 23(2–3):83–104
26. Geissdoerfer M, Savaget P, Bocken NM, Hultink EJ (2017) The circular economy–a new sustainability paradigm? J Clean Prod 143:757–768
27. Ghobakhloo M (2018) The future of manufacturing industry: a strategic roadmap toward Industry 4.0. J Manuf Technol Manag 29(6):910–936
28. Ghobakhloo M (2020) Industry 4.0, digitization, and opportunities for sustainability. J Cleaner Prod 252:119869
29. Gretzel U, Sigala M, Xiang Z, Koo C (2015) Smart tourism: foundations and developments. Electron Mark 25:179–188
30. Hashem IAT, Chang V, Anuar NB, Adewole K, Yaqoob I, Gani A, ... Chiroma H (2016) The role of big data in smart city. Int J Inf Manag 36(5):748–758
31. Helu M, Libes D, Lunell J, Lyons K, Moris KC (2016) Enabling smart manufacturing technologies for decision-making support. Proceedings of the ASME 2016 international design engineering technical conferences & computers and information in engineering conference IDETC/CIE, Charlotte, NC. 1–10. August 21–24
32. Ivanov D, Dolgui A, Sokolov B (2019) The impact of digital technology and Industry 4.0 on the ripple effect and supply chain risk analytics. Int J Prod Res 57(3):829–846
33. Jones MV, Coviello N, Tang YK (2011) International entrepreneurship research (1989–2009): a domain ontology and thematic analysis. J Bus Ventur 26(6):632–659
34. Kang Q, Li H, Cheng Y, Kraus S (2021) Entrepreneurial ecosystems: analysing the status quo. Knowl Manag Res Pract 19(1):8–20
35. Kang HS, Lee JY, Choi S, Kim H, Park JH, Son JY, ... Noh SD (2016) Smart manufacturing: Past research, present findings, and future directions. Int J Precision Eng Manuf Green Technol 3(1):111–128
36. Kruggel A, Tiberius V, Fabro M (2020) Corporate citizenship: structuring the research field. Sustainability 12(13):5289
37. Kusiak A (2018) Smart manufacturing. Int J Prod Res 56(1–2):508–517
38. Lee I, Lee K (2015) The internet of things (IoT): applications, investments, and challenges for enterprises. Bus Horiz 58(4):431–440
39. Li G, Hou Y, Wu A (2017) Fourth Industrial revolution: technological drivers, impacts and coping methods. Chin Geogra Sci 27(4):626–637
40. Liao Y, Deschamps F, Loures EDFR, Ramos LFP (2017) Past, present and future of Industry 4.0-a systematic literature review and research agenda proposal. Int J prod res 55(12):3609–3629
41. Liu Y, Xu X (2017) Industry 4.0 and cloud manufacturing: a comparative analysis. J Manuf Sci Eng 139(3)

42. Lu Y, Morris KC, Frechette S (2016) Current standards landscape for smart manufacturing systems. Nat Inst Standards Technol NISTIR 8107:39
43. Luther L, Tiberius V, Brem A (2020) User Experience (UX) in business, management, and psychology: a bibliometric mapping of the current state of research. Multimodal Technol Interaction 4(2):18
44. Machado CG, Winroth MP, Ribeiro da Silva EHD (2020) Sustainable manufacturing in Industry 4.0: an emerging research agenda. Int J Prod Res 58(5):1462–1484
45. Macke N, Rulhoff S, Stjepandic J (2016) Advances in smart manufacturing change management. In: ISPE TE, pp 318–327
46. Malaga A, Vinodh S (2021) Benchmarking smart manufacturing drivers using Grey TOPSIS and COPRAS-G approaches. Benchmarking Int J 28(10):2916–2951
47. Mas-Tur A, Kraus S, Brandtner M, Ewert R, Kürsten W (2020) Advances in management research: a bibliometric overview of the review of managerial science. RMS 14(5):933–958
48. Merediz-Solà I, Bariviera AF (2019) A bibliometric analysis of bitcoin scientific production. Res Int Bus Financ 50:294–305
49. Mittal S, Khan MA, Romero D, Wuest T (2019) Smart manufacturing: Characteristics, technologies and enabling factors. Proceedings of the institution of mechanical engineers, Part B: J Eng Manuf 233(5):1342–1361
50. NIST (2014) Smart manufacturing operations planning and control program
51. Ngai EWT, Moon KK, Riggins FJ, Candace YY (2008) RFID research: an academic literature review (1995–2005) and future research directions. Int J Prod Econ 112(2):510–520
52. Noyons EC, Moed HF, Luwel M (1999) Combining mapping and citation analysis for evaluative bibliometric purposes: a bibliometric study. J Amer Soc Inf Sci 50(2):115–131
53. O'Donovan P, Bruton K, O'Sullivan DT (2016) Case study: the implementation of a data-driven industrial analytics methodology and platform for smart manufacturing. Int J Prognostics Health Manag 7(3):1–21
54. Pan Y (2016) Heading toward artificial intelligence 2.0. Engineering 2(4):409–413
55. Pranckutė R (2021) Web of Science (WoS) and Scopus: The titans of bibliographic information in today's academic world. Publications 9(1):12
56. Qi Q, Tao F (2018) Digital twin and big data towards smart manufacturing and industry 4.0: 360 degree comparison. IEEE Access 6:3585–3593
57. Radziwon A, Bilberg A, Bogers M, Madsen ES (2014) The smart factory: exploring adaptive and flexible manufacturing solutions. Procedia Eng 69:1184–1190
58. Raza SA, Govindaluri SM, Bhutta MK (2022) Research themes in machine learning applications in supply chain management using bibliometric analysis tools. Benchmarking Int J. https://doi.org/10.1108/BIJ-12-2021-0755
59. Rossetto DE, Bernardes RC, Borini FM, Gattaz CC (2018) Structure and evolution of innovation research in the last 60 years: review and future trends in the field of business through the citations and co-citations analysis. Scientometrics 115(3):1329–1363
60. Saberi S, Kouhizadeh M, Sarkis J, Shen L (2019) Blockchain technology and its relationships to sustainable supply chain management. Int J Prod Res 57(7):2117–2135
61. Shafiq SI, Sanin C, Toro C, Szczerbicki E (2015) Virtual engineering object (VEO): Toward experience-based design and manufacturing for industry 4.0. Cybern Syst 46(1–2):35–50
62. Silva JTM, Ablanedo-Rosas JH, Rossetto DE (2019) A longitudinal literature network review of contributions made to the academy over the past 55 years of the IJPR. Int J Prod Res 57(15–16):4627–4653
63. Stock T, Seliger G (2016) Opportunities of sustainable manufacturing in industry 4.0. Procedia CIRP 40:536–541
64. Tao F, Cheng Y, Da Xu L, Zhang L, Li BH (2014) CCIoT-CMfg: cloud computing and internet of things-based cloud manufacturing service system. IEEE Trans Industr Inf 10(2):1435–1442
65. Tao F, Qi Q, Wang L, Nee AYC (2019) Digital twins and cyber–physical systems toward smart manufacturing and industry 4.0: correlation and comparison. Engineering 5(4):653–661
66. Thoben KD, Wiesner S, Wuest T (2017) "Industrie 4.0" and smart manufacturing-a review of research issues and application examples. Int J Autom Technol 11(1), 4–16

67. Tilling M (2004) "The triple bottom line: does it all add up?", henriques and Richardson. Soc Environ Account J 24(2):32–33
68. Tiwari S (2015) Framework for adopting sustainability in the supply chain. Int J Autom Logist 1(3):256–272
69. Tiwari S, Dubey R, Tripathi N (2011) The journey of lean. Indian J Commer Manag Studies 2(2):200–210
70. Tiwari S, Tripathi N (2012) Lean manufacturing practices and firms performance measurement—a review paper. J Supply Chain Manag Syst 1(1):44
71. Tiwari S, Bahuguna PC, Walker J (2022) Industry 5.0: a macroperspective approach. in handbook of research on innovative management using AI in industry 5.0, pp. 59–73. IGI Global
72. Tiwari S (2021) Supply chain integration and Industry 4.0: a systematic literature review. Benchmarking Int J 28(3):990–1030
73. Tiwari S, Bahuguna PC, Srivastava R (2022). Smart manufacturing and sustainability: a bibliometric analysis. Benchmarking Int J. https://doi.org/10.1108/BIJ-04-2022-0238
74. Tiwari S (2022) Supply chain innovation in the era of industry 4.0. In: Handbook of research on supply chain resiliency, efficiency, and visibility in the post-pandemic era, pp 40–60. IGI Global
75. Tiwari S, Srivastava R (2022) Cyber security trend analysis: an Indian perspective. In: Cross-industry applications of cyber security frameworks, pp 1–14. IGI Global.
76. Tiwari S, Raju TB (2022) Management of digital innovation. In: Promoting inclusivity and diversity through internet of things in organizational settings, pp 128–149. IGI Global
77. Vanhala M, Lu C, Peltonen J, Sundqvist S, Nummenmaa J, Järvelin K (2020) The usage of large data sets in online consumer behaviour: a bibliometric and computational text-mining–driven analysis of previous research. J Bus Res 106:46–59
78. Wang L (2019) From intelligence science to intelligent manufacturing. Engineering 5(4):615–618
79. Wang J, Gao RX (2022) Innovative smart scheduling and predictive maintenance techniques. In design and operation of production networks for mass personalization in theeEra of cloud technology (pp. 181–207). Elsevier
80. Williams R, Edge D (1996) The social shaping of technology. Res Policy 25(6):865–899
81. Xu LD, Xu EL, Li L (2018) Industry 4.0: state of the art and future trends. Int J Prod Res 56(8):2941–2962
82. Yadav G, Luthra S, Jakhar SK, Mangla SK, Rai DP (2020) A framework to overcome sustainable supply chain challenges through solution measures of industry 4.0 and circular economy: an automotive case. J Cleaner Prod 254:120112
83. Zhang L, Luo Y, Tao F, Li BH, Ren L, Zhang X, ... Liu Y (2014) Cloud manufacturing: a new manufacturing paradigm. Enterp Inf Syst 8(2):167–187
84. Zhong RY, Xu C, Chen C, Huang GQ (2017) Big data analytics for physical internet-based intelligent manufacturing shop floors. Int J Prod Res 55(9):2610–2621
85. Zhong RY, Xu X, Klotz E, Newman ST (2017b). Intelligent manufacturing in the context of industry 4.0: a review. Engineering 3(5):616–630
86. Zhou J, Li P, Zhou Y, Wang B, Zang J, Meng L (2018) Toward new-generation intelligent manufacturing. Engineering 4(1):11–20
87. Zhou J, Zhou Y, Wang B, Zang J (2019) Human–cyber–physical systems (HCPSs) in the context of new-generation intelligent manufacturing. Engineering 5(4):624–636

Performance Metrics in Digital Supply Chain Paradigm

R. Prashanna Rangan, R. Kumanan, K. Ram Prasad, and M. Nishal

Abstract Globalisation in corelation with technology development has recently made it possible to compete in business and prosper. Digital technology must now be included by companies if they are to remain sustainable and globally competitive. The fourth industrial revolution has altered how individuals interact with their surroundings and communicate. The supply chain has begun to be digitalised in order to accommodate digital goods and services as internal business processes change so quickly. Innovative strategies, including technologies for digital transformation, are required to obtain an advantage in the digital supply chain. Current methods for global supply chain management and logistics for moving, storing and managing cargo are not suitable for the future from an economic, environmental or social standpoint. To assess how effectively a company is performing in achieving its goals, supply chain metrics and performance measures obtained through benchmarking are essential. Monitoring a company's operations, performance and behaviour is made possible via performance metrics. They ensure that the supply chain is controlled and operating efficiently by enabling visibility into key operations. This chapter provides insights into the performance metrics which control the supply chain activities internally and externally. The Key Performance Indicators known as KPI's which drive the organisation towards its intended results have been listed, and their significance towards the digital supply chain has been elaborated in detail. The measurements can be used to determine inventory accuracy, turnover and inventory-to-sales ratios. Therefore, by monitoring supply chain indicators, a business may quickly pinpoint areas that require improvement or where supply chain effectiveness may be increased. To maximise the gains of implementing a Digital Supply Chain (DSC), it is crucial to identify all of the potential opportunities, pain spots, difficulties and development chances beforehand.

Keywords Digital Supply Chain (DSC) · Performance metrics · DSC assessment · Industry 4.0

R. P. Rangan (✉) · R. Kumanan · K. R. Prasad · M. Nishal
Department of Mechanical Engineering, Sri Venkateswara College of Engineering, Sriperumbudur, India
e-mail: prashanna098@gmail.com

K E K Vimal et al. (eds.), *Industry 4.0 Technologies: Sustainable Manufacturing Supply Chains*, Environmental Footprints and Eco-design of Products and Processes,
https://doi.org/10.1007/978-981-99-4819-2_11

1 Introduction

A web of all the personnel, departments, agencies, businesses, assets, activities and technological improvements in manufacturing and distribution of goods is generally termed supply chain. A digital supply chain [1–3], on the other hand, gives better perceptions of the roles of each stakeholder across the chain by utilising the latest technologies. Having used these cutting-edge technologies every member of supply chain can make healthier verdicts about the foundations of resources they require and their subsequent demand. Supply chain metrics and performance measures via benchmarking [4] are necessary for determining how well a company is doing in reaching its objectives. By providing visibility into crucial operations, they guarantee that the supply chain is controlled and running effectively. As a result, a company can easily identify areas that need improvement or where supply chain effectiveness might be raised by tracking supply chain indicators. By improving their awareness of their supply chains, businesses may increase their flexibility and resilience. They may also approach challenges [5] more proactively.

Supply Chain Performance Measurement (SCPM) is a quantifying process that evaluates the productivity and overall efficacy of the supply chain in real-time which eventually reflects the key areas for improvement and enhancement as the SCPM explicitly tells whether the supply chain has improved or degraded. Lack of adequate performance measurement will hinder efficient Supply Chain Management (SCM) and a good performance measurement will lead to increased visibility of measures thus pointing out the key improvement areas and paving a way for structured strategic and tactical decision-making to resolve and eliminate the identified issues by offering a closed loop system that takes care of the SCM. Inclusiveness, universality, measurability and consistency are the main traits for implementing an effective SCPM in correlation with non-conflicting, clearly defined metrics for performance calculations. Earlier only financial accounting metrics were used for SCPM which started to migrate slowly into the balanced integrated system that included non-financial performance measurement and with further advancements and necessities more criteria came into play, viz, dimensions, functions, perspectives, processes, management levels, strategy, overall efficiency, linkages, etc. The seven dimensions of a Digital Supply administration for enabling better management are elaborated as Digital performance measurement, Digital IT and Technology, Digital Human Resource, Digital Suppliers, Digital Manufacturing Systems, Digital Inventory Logistics and Digital Customers. Overall architecture and essential constructs of DSC includes SC Management components, SC Management processes, SC Network Structure and SC Flows. A Digital Supply Chain Model involving Industry 4.0 has a virtual value chain that links with the Digital and Physical world that owes several processes involving the digital and physical SCM elements viz technology and equipment's combined to work in a cohesive fashion.

The various challenges, pain points and opportunities and potential scope for development for implementing a Digital Supply Chain (DSC) have to be identified prior to the implementation in order to get the maximum outcome of it [7]. A detailed study [8] is made on acquiring the approaches in obtaining the benefits of digital supply chain has been made and mapped their transformation and performances [9]. The need for the migration [10] from supply chain to a DSC and its impacts on the organisational performance with respect to 15 critical factors for dependency and highest driving power factors. An analytic hierarchy process [11] has been introduced for measuring the green supply chain performance. Incorporating the concepts of Industry 4.0 with the DSC will yield more benefits and its dynamic capabilities [12] were elaborated with a theoretical framework. Intrusion of Industry 4.0 and digital twin in supply chain activities, and an attempt to merge both the concepts were put forth for managing the disruption vulnerabilities and to optimise the flexibility of DSC [13].

A Knowledge-Based System [14] (KBS) involving fuzzy implementation for enhancing the outcome of a DSC was proposed which evaluated the readiness factor of the DSC which enables the managers to get a strategy to realise digitalisation utterly in their SC operations. The advantages and impact of using the DSC have been tested in real-time with an empirical study on manufacturing [15] and food and beverage [16] industries for analysing the performance of DSC. Seventeen use cases for a DSC management for improving the automobile industry [17] and transforming it into demand-sensitive networks were assessed for their value and applicability. An optimisation framework [18] involving the DMAIC technique has been proposed for improvising the productivity [19] of the supply chain adjacent to a case study has been carried out. Blockchain has completely changed the way goods are tracked throughout the supply chain since it is so efficient and transparent. Blockchain is also employed to guarantee the validity and origin of products. This digital ledger technology has made it possible for various company areas. Blockchain technology combined with QR code application has made it possible to build low-cost methods of preventing the sale of fake medicines in developing nations.

Evaluation of DSC using SCOR metrics by utilising a Pythagorean fuzzy AHP method [20] and Adaptive Network-based Fuzzy Inference System [21] (ANFIS), fuzzy TOPSIS [22] has been carried out that resulted in improved precision of extrapolation, understanding capability based on chronological trend, fitness to encourage decision-making in ambiguity, clearer interpretability of outcomes, among others. The fuzzy TOPSIS [22] in addition is applied for ranking the metrics in terms of flexibility and responsiveness in an IoT-based SCPM framework for agriculture supply chain. A DSC's performance [23, 24] could be improved by incorporating the intelligent autonomous vehicles [25] and a framework for such integration has been put forth for developing a sustainable network of DSC. System of measurement of performance [26] is vital to the triumph of an organisation whose key performances are assessed directly using these metrics. Significant achievement factors help enhancing their performances and hence it is imperative for an organisation to examine them continuously.

Though the advancements are there, the conventional approach to SCPM limits and hinders the overall effectiveness of the SCPM as it is inflexible and a multidisciplinary approach incorporating systems thinking, strategic planning and optimisation methods could address such limitations. Rapid advancements in artificial intelligence, blockchain, cybersecurity, condition monitoring, robotics and industrial IoT are the key enablers for recital dimension in a DSC, that blends the aforementioned technologies to leverage an automated platform that encompasses the production at its optimum level without flaws.

2 Performance Metrics

Performance metrics [19] enable the monitoring of a company's operations, performance and behaviour. In order to facilitate the construction of a foundation in aid of the achievement of overarching enterprise intentions, this should take the shape of quantities of the vital information within a range. Supply chain metrics are characterised by setting predetermined characteristics that are used to quantify and characterise supply chain operations. Inventory accuracy, turnover and inventory-to-sales measurements can be based on the measurements. Performance indicators are essential because they provide your business with important data. These indicators' information can help you grow your business and increase profitability. They also facilitate the execution of strategies for reaching various objectives. Supply chain metrics must be taken into consideration in the company's crucial fulfilment and logistics plan for a number of reasons. First, research indicates that by 2027, the market for supply chain analytics would be worth $16.82 billion. This is because by using analytics to inform supply chain decisions, businesses may boost their tactical, strategic and operational effectiveness.

Giving organisations meaningful data that will enable them to better meet customer needs and realise their strategic goals is the primary purpose of performance measurement. Supply chain performance measurement is used to weigh the efficacy and effectiveness of organisational formations, practices and reserves for the entire supply chain as well as for a single enterprise. It provides some context for getting insight into the overall mechanism, influences behaviour and provides information on how stakeholders and supply chain actors are performing. A critical management activity is the development and use of performance measurements. The purposes of lucidity and a collective grasp of the entire supply chain are similarly backed by the application of performance-measuring tools [23]. SCOR performance attributes of a DSC is split into two components with the first component focussing on customer reliability, flexibility and responsiveness whereas the second component focusses on the internal cost and assets of a DSC.

2.1 *Internal Supply Chain Performance Measurement*

Performance measurements of an internal supply chain primarily focuses on preparedness, replenishment rate and on-time delivery. These internal supply chain performance monitoring tools are essential for identifying problems, facilitating internal communication within a company and assisting with decision-making. They assess not only the efficiency of operations but also the outcomes and goals of the firms. Conventional performance assessment techniques have also come under fire for being too short-term and finance-focused, ignoring strategic relevance, placing a narrow emphasis internally, avoiding overall improvements, utilising inconsistent measurements and expressing performance in numerical terms. When these functions of core performing assessment and the accompanying critiques are considered, it becomes evident that these inner performance evaluation systems cannot be translated to peripheral performance assessment practices that would measure the unified supply chain. As a result, methods for external supply chain performance monitoring that reach outside organisations' and their core competencies' boundaries must now be developed.

2.2 External Supply Chain Performance Metrics

Systems for evaluating performance usually place an emphasis on local optimisation and lack system thinking. They typically don't use balanced methods that take into account both financial and non-financial facts. When these challenges are considered, combined with the reality that the majority of forms are realising the need to manage supply chain activities, it pushes the need to assess performance assessment tools in a supply chain and utilise them for supply chain operations. The supply chain environment's existing performance measuring techniques usually fall short of expectations because of the multiple vertical requirements.

2.3 Requirements for Performance Measurement Metrices

- Genuine expression of organisational performance.
- Based on corporate goals and strategy.
- Ensure that both financial and non-financial metrics are in balance.
- Decision-making and control at the strategic, tactical and operational levels.
- Be comparable to other performance metrics employed by organisations with similar missions.
- The goal, data collecting and calculation techniques, update and monitoring mechanisms and associated processes should all be clearly stated.
- Vary between organisational locations and be managed by the organisational unit that was examined.
- Permit target-setting, aggregation and disaggregation.
- Enable weighting and prioritisation.
- Promote integration.
- Do not overlap.
- Manage intricate overhead structures.

- Be straightforward and simple to use; ratios are preferable to absolute figures.
- To be more actionable, be specific and non-financial rather than aggregate and financial.
- Be decided upon after consultation with all parties concerned and cater to the needs of individuals at all levels (not only upper management).
- Take a proactive stance to facilitate quick feedback and ongoing progress.
- Be reliable and true.
- Be transparent and coherent.
- Rely on your experience.
- Make room for organisational learning, which involves testing, reviewing, rewriting and refining.
- Provide the fewest possible indicators with the best possible accuracy at the lowest possible price.
- Be able to quantify business excellence and other concepts like collaboration, cooperation, responsiveness, adaptability and data efficiency.

3 Classification of Performance Metrics

Based on the literature review that has been performed 20 key performance metrics were identified and listed in the Table 1 along with the potential area of assessment for DSC.

3.1 Delivery Time

Key performance indicator (KPI) is a measure of a specified parameter for a specified objective over a period of time. In supply chain, time taken for delivery is a key KPI which has been given primary emphasis to enhance customer assistance. It provides an estimate of how long it will take a product to travel from the place of dispatch to the customer's end. Prior to delivery, careful preparation and packaging are required for the order, failing which the customer may feel dissatisfied and will lead losing of customers. Reducing this supply chain management KPI makes sense in favour of improving the accuracy of the delivery information that customers receive. For instance, stating that the consignment would arrive in 2–3 working days is preferable for stating that it will take 5–8 business days.

Furthermore, it would greatly improve your service if you could provide the time. To expedite delivery, you can also offer specialised delivery services. Over time, you can monitor how this affects customer satisfaction. Better yet, add supply delivery indicators to your supply chain-focused performance dashboard to help you keep closer tabs on it.

Table 1 Performance Metrics with area of assessment for DSC

S. no	Performance metrics	Area of assessment in DSC
1	Delivery time	IOT, data analysts
2	Time cycle in cash-to-cash	Forecasting, bigdata analytics
3	Inventory movement	Automated product delivery system, Robotics
4	Gross margin return on investment (GMROI)	Big data analytics, ERP
5	Absolute order rate	Warehousing
6	Day's scales outstanding (DSO)	ERP
7	Cycle time for the supply chain	Robotics
8	Customer order cycle time	IIOT, blockchain
9	Fill rate	Forecasting
10	Warehousing costs	Forecasting
11	Cost of shipping per unit	Predictive analysis
12	Average delivery time	Forecasting
13	Cycle time for pick and pack	Inventory
14	Ratio of inventory to sales	Inventory
15	Inventory velocity	Logistics
16	Return ground	ERP, blockchain
17	Costs in a supply chain	ERP
18	Sales versus costs in supply chain	Purchasing, demand planning
19	On-time shipping	IOT, blockchain
20	Use of packaging material	Big data analytics

3.2 *Time Cycle in Cash-to-Cash*

This invaluable supply chain statistic allows us to estimate how long it will take to convert your resources into actual cash flows. In the cash-to-cash time cycle, days of payables (DOP), days of inventory (DOI) and days of receivables (DOR) are three essential ratios (DOR). KPI demonstrates how long it takes for a company to get money from its clients after having already paid its suppliers. The sharper the transfer cycle, the effective and you can utilise this unique supply chain data to decide how to manage your business more successfully.

3.3 Inventory Movement

One of the most important KPIs currently accessible is the rate at which inventories are sold and replenishment period for the same. This is a very good indication of effective production planning, marketing and sales management and process strategy. It is necessary to build a direct reporting to management in case of delays and take proper action to increase on-time shipment rate for a period of time by assessing on-time shipment frequency. As a result, brand authority will increase, thereby improving the bottom line.

3.4 Gross Margin Return on Investment (GMROI)

Although one of the most important indicators you can use in your company is GMROI, it might be a little intimidating. It demonstrates how diligently your inventory works to bring in money for you. GMROI is a vital metric for preserving your cash flow. Although it's important for many businesses to protect their margins, bear in mind that you can live without money for a while. You cannot live without financial flow.

Even though some of your inventory moves slowly, it still makes good money. Some enterprises might change hands quickly but with little profit. These two scenarios will produce GMROIs that are reasonable to high. When a product has a low GMROI, it either moves too slowly, has insufficient margin, or both. Fortunately, you can assess the performance of a product using GMROI calculation.

3.5 Absolute Order Rate

This peculiar data is one of the highly crucial supply chain KPIs for companies across numerous industries. Your capacity to complete orders without incident is measured by your perfect order rate, which will eventually assist you deal with problems like errors, harms, interruptions and inventory failures. The optimal order frequency is a KPI that requires improvement because it directly affects your customer loyalty and retention rates.

3.6 Day's Sales Outstanding (DSO)

The KPI for the day's outstanding sales measures how quickly you can make money or collect payments from customers. A corporation that pays its accounts receivable in fewer days has a low or healthy DSO figure. A company with a higher DSO level sells its products on credit to customers and takes longer to collect payment, which can restrict cash flow and lower total profits. Making frequent calculations will help you collect money more swiftly and efficiently, thus improving your bottom line.

3.7 Cycle Time for the Supply Chain

A comprehensive statistic for measure of cycle time in supply chain is the duration taken for a customer order to be completed when all inventory levels are zero. In every phase of supply chain cycle, the highest lag periods are added up to produce this statistic. This number is an excellent way to measure the effectiveness of supply chain. A process that has a shorter cycle is more adaptable, nimble and receptive to external stimuli.

3.8 Customer Order Cycle Time

The time taken from the point of order placed by the customer to the point of order delivered is the customer order cycle time. While a short wait time excites customers and demonstrates your supply chain's high efficacy, the customer order cycle time should be kept as low as feasible. Customer order cycle time can be drastically lowered by automatically routing each order to the fulfilment plant that is closest to its intended destination. Shortening the cycle might be achieved by accelerating backend operations, streamlining SOPs for warehouse management and keeping track on shipping carrier performance.

3.9 Fill Rate

The percentage of orders from the stock available that can be shipped to satisfy the client demand under the condition that there is no deprivation of backorders or lost sales is termed as the fill rate. Since it reveals the sales for which better inventory management could lead to better service, understanding your fill rate is crucial. Transparency in inventory data and its ease of access to all supply chain partners can help teams to ship correct, complete and on-time orders. Subsequently, sales and customer satisfaction will be improved as the teams will be better equipped with inventory data and its readiness to the chain.

Research suggests that enhancing the correlation between a retailer and a supplier can boost fill rates by 80%. improving responses to demand surges, accelerating conversations about pricing changes, streamlining order management processes and changing sales incentives.

3.10 Warehousing Costs

The cost of storage is the next statistic on our list of supply chain indicators. Controlling your inventory's time and space while properly allocating expenditures will result in a solid supply chain. Even though these expenses differ from warehouse to warehouse, it's still crucial to keep an eye on this signal and examine it on a regular basis in order to spot possibilities and cut back on unnecessary spending. Managing the warehouse facility involves paying for a number of expenses, including staff, warehouse rent, electricity, equipment, supplies and an information processing system, as well as purchasing and storing the items.

Being well informed on all of the operations at the warehouse inventory and how precisely it performs is the first step in keeping costs as low as is practical. This will provide you a greater opportunity to cut back on wasteful spending, use operations management techniques and make the required adjustments. Also, you will be able to trust your statements and make quicker, way more precise business determinations if you regularly collect your data utilising a reliable online reporting tool.

3.11 Cost of Shipping per Unit

For businesses that are planning to expand, the freight cost per unit is one of the most crucial supply chain KPIs. They provide a clear sense of how inexpensively goods can be sent and are much important to achieve long-term benefits.

The specific function of the supply chain system of measurement console is to compute the whole cargo expenses and divide them by the volume of products transported. You are allowed to estimate this measure employing any suitable unit for your company. But, any drawn-out or inadequately prepared procedure that will lead to loss of time, money and loyalty can be avoided by focusing on these KPIs.

3.12 Average Delivery Time

The time it takes for a product to go from your fulfilment centre to a customer's door is the typical delivery time. This is a crucial metric to monitor because it reveals how swiftly a business ships its goods. Online shoppers are growing irritated with the processing time for their orders due to the popularity of 2-day delivery. Although expedited shipping reduces the amount of time between orders, smaller businesses might find it to be a costly long-term solution. As a result, many e-commerce companies decide to shorten the typical delivery period by carefully positioning some of their inventory close to the target clients.

3.13 Cycle Time for Pick and Pack

By dissecting your supply chain cycle into distinct segments, this supply chain administration assessment will provide you a thorough understanding of how effective (or unproductive) the whole supply chain cycle is. For instance, each KPI measures the amount of time it takes an employee to pick and package an item after selecting it off the shelf.

Once you've established your goals and begun monitoring the progress of the supply chain cycle, it will be obvious in which the impediments or flaws in your system are. In order to stop these problems in their tracks and reduce your overall cycle times, you might take specific action.

3.14 Ratio of Inventory to Sales

Since cargo is one of the most crucial tools in your supply chain, the stockpile-to-sales ratio is one of the important supply chain metrics that must be monitored. The ratio between the amount of items available for purchase and the amount that are actually sold is established by this statistic. Also, it will demonstrate how well your business adapts to unforeseen circumstances and assist you in changing your items to produce high margins.

The key idea here is that maintaining a healthy ratio requires understanding how to balance it correctly. It would make sense to avoid having a high proportion because it can lower your inventory turnover rates. Finding the ideal balance in this situation is crucial. You may create an interactive inventory KPI with a modern dashboard builder that will automatically refresh the data and allow you to track performance in real time. Also, you can adjust your future plans to ensure that the ratio is appropriate for your particular organisation.

3.15 Inventory Velocity

One of the most crucial supply chain KPIs for displaying visually how much inventory is anticipated to be used during the upcoming month or quarter is inventory velocity. Inventory Velocity is a supply chain statistic that will assist you in optimising your inventory levels, increasing your chances of satisfying consumer demand and avoiding losses brought on by excessive stock levels. It is computed by taking the opening stock into account and subtracting the anticipated sales for the subsequent time frame.

3.16 Return Ground

Understanding the various reasons why customers and clients return their goods is essential to an eCommerce fulfilment company's continued success. The supply chain metrics for return reasons give this data which will help to figure out the causes for return, examine vulnerabilities in the chain and end up making the sorts of advancements that will substantially enhance not only your notoriety but also your overarching level of service. You also greatly improve your prospects of lowering returns, increasing earnings and enhancing cash flow by acquiring this level of information.

3.17 Costs in a Supply Chain

Costs associated with every stage of supply chain is the primary KPI for supply chain management. These costs, which show how productive each department of the company is, may cover things like planning, team management, sourcing, delivery, etc. Of course, any business must look for ways to increase earnings, and one common strategy is decreasing costs. With this approach, the company can assess its potential for growth without also needing to increase sales. It's vital to take into account how the cost reduction would impact the entire supply chain.

3.18 Sales Versus Costs in Supply Chain

Further cost analysis that is connected to sales is also included in our list of supply chain KPIs and indicators. Basically, by measuring your supply chain expenses as a % of sales, this indicator will demonstrate the amount you're expending in comparison to your consumption expenditure. With these supply chain management indicators, you can do a thorough cost analysis and set up procedures for potential cost savings. Indeed, cost reduction is a key component of supply chain optimisation. Yet, as we've already stated, it's critical to minimise costs when they make sense rather than just simply lower the statistics. There is a simple rationale involved as costs are reduced in one part of your supply chain it tends to increase in another, the entire process becomes pointless.

3.19 On-Time Shipping

Using a specialist KPI tool, on-time shipping was created as a wonderful indicator of how long it might take you to dispatch a specific kind of order to a customer, purchaser or partner. By using this KPI, you can determine the point of reference distribution time for every single product, which will help you streamline your delivery and shipping procedures and speed up customer satisfaction.

3.20 *Use of Packing Material*

The use of packing materials is the next KPI on our list for the supply chain. Inefficient packing techniques waste money and create extra rubbish for your business. With the help of this scannable supply chain KPI, you can securely monitor the quantity of packaging resources you are utilising for each field in your pick-up and packing procedure. Define a goal or benchmark for your package control to keep track of it and prevent any potential issues (in this case, 300 g). By examining your packaging practices before costs escalate, you may save money and your brand's reputation will rise as a result of your newfound sustainability.

4 Conclusion and Scope for Future Work

The effects of the industrial revolution, communication technologies, Internet of things, architecture of a cyber-physical system for fabrication logistics and supply chain functions have hastened the innovations needed for the digitalisation of companies. Managing the supply chain from start to finish has always been a time-consuming and inefficient process for manufacturers, requiring more time and manpower. Automakers had already benefitted immensely from more promising technologies for requirement forecasting and logistics management—the very first and last elements of their supply chains—but monitoring and measuring the manufacturing production performance across the supply chain has remained trapped in the epoch of whiteboards, clipboards, spreadsheets and manually generated documentations. This study about performance metrics in this chapter could enable industry practitioners and researchers to have a quantifiable measurement over the supply chain activities. They will be pushing hard with all supply chain actors to have an end-to-end track of all its activities otherwise, which could now be reduced throughout the supply chain. The KPI among each process and within the processes could now be earmarked which makes SC actors to have an easier traceability of their performance metrics. Future researchers who work in DSC can go for a deeper insight into the metrics listed here, and their correspondence to each of the technological aspects could further facilitate organisations to standardise the digitisation process across supply chain and thereby providing prominent satisfaction to its customer even without altering their system to a greater extent. Firms should also understand that maximising the paybacks of DSC largely depends on utilising the above-mentioned performance metrics to its fullest extent; hence, it is ideal to have a clear understanding of the activities they do and identifying their related KPI which has the potential to drive them to their planned objectives.

Declarations

Availability of Data and Materials Not applicable.

Competing Interests The authors declare that they have no known competing financial interests or personal relationships that could have appeared to influence the work reported in this chapter.

References

1. Büyüközkan G, Göçer F (2018) Digital supply chain: Literature review and a proposed framework for future research. Comput Ind 97:157–177. https://doi.org/10.1016/j.compind.2018.02.010
2. Agrawal P, Narain R (2018) Digital supply chain management: an overview. In: IOP conference series: materials science and engineering, vol 455, no 1. https://doi.org/10.1088/1757-899X/455/1/012074
3. Iddris F (2018) Digital supply chain: survey of the literature. Int J Bus Res Manag (IJBRM) 9(1):47–61, [Online]. Available: http://urn.kb.se/resolve?urn=urn:nbn:se:hh:diva-37526
4. Wong WP, Wong KY (2008) A review on benchmarking of supply chain performance measures. Supply Chain Perform Measur 15(1):25–51. https://doi.org/10.1108/14635770810854335
5. Ageron B, Bentahar O, Gunasekaran A (2020) Digital supply chain: challenges and future directions. Supply Chain Forum 21(3):133–138. https://doi.org/10.1080/16258312.2020.1816361
6. Garay-Rondero CL, Martinez-Flores JL, Smith NR, Caballero Morales SO, Aldrette-Malacara A (2020) Digital supply chain model in Industry 4.0. J Manuf Technol Manag 31(5):887–933, https://doi.org/10.1108/JMTM-08-2018-0280
7. Queiroz MM, Pereira SCF, Telles R, Machado MC (2019) Industry 4.0 and digital supply chain capabilities: a framework for understanding digitalisation challenges and opportunities. Benchmarking 28(5):1761–1782. https://doi.org/10.1108/BIJ-12-2018-0435
8. Nasiri M, Ukko J, Saunila M, Rantala T (2020) Managing the digital supply chain: the role of smart technologies. Technovation 96–97. https://doi.org/10.1016/j.technovation.2020.102121
9. Gawankar SA, Gunasekaran A, Kamble S (2020) A study on investments in the big data-driven supply chain, performance measures and organisational performance in Indian retail 4.0 context. Int J Prod Res 58(5):1574–1593. https://doi.org/10.1080/00207543.2019.1668070
10. Ahmed Khan S, Kusi-Sarpong S, Gupta H, Kow Arhin F, Nguseer Lawal J, Mehmood Hassan S (2021) Critical factors of digital supply chains for organizational performance improvement. IEEE Trans Eng Manag, https://doi.org/10.1109/TEM.2021.3052239
11. Dey PK, Cheffi W (2013) Green supply chain performance measurement using the analytic hierarchy process: a comparative analysis of manufacturing organisations. Prod Planning Control 24(8–9):702–720. https://doi.org/10.1080/09537287.2012.666859
12. Gupta S, Modgil S, Gunasekaran A, Bag S (2020) Dynamic capabilities and institutional theories for Industry 4.0 and digital supply chain. Supply Chain Forum 21(3):139–157. https://doi.org/10.1080/16258312.2020.1757369
13. Ivanov D, Dolgui A (2021) A digital supply chain twin for managing the disruption risks and resilience in the era of Industry 4.0. Prod Planning Control 32(9):775–788. https://doi.org/10.1080/09537287.2020.1768450
14. Khan SA, Naim I, Kusi-Sarpong S, Gupta H, Idrisi AR (2021) A knowledge-based experts' system for evaluation of digital supply chain readiness. Knowl Based Syst 228. https://doi.org/10.1016/j.knosys.2021.107262
15. Lee KL, Azmi NAN, Hanaysha JR, Alzoubi HM, Alshurideh MT (2022) The effect of digital supply chain on organizational performance: an empirical study in Malaysia manufacturing

industry. Uncertain Supply Chain Manag 10(2):495–510. https://doi.org/10.5267/j.uscm.2021.12.002
16. Saryatmo MA, Sukhotu V (2021) The influence of the digital supply chain on operational performance: a study of the food and beverage industry in Indonesia. Sustainability (Switzerland) 13(9). https://doi.org/10.3390/su13095109
17. Farahani P, Meier C, Wilke J (2016) Digital supply chain management agenda for the automotive supplier industry. In: Shaping the digital enterprise: trends and use cases in digital innovation and transformation, Springer International Publishing, pp 157–172. https://doi.org/10.1007/978-3-319-40967-2_8
18. Farsi M et al (2020) An optimisation framework for improving supply chain performance: case study of a bespoke service provider. Procedia Manuf 49:185–192. https://doi.org/10.1016/j.promfg.2020.07.017
19. Kamble SS, Gunasekaran A (2020) Big data-driven supply chain performance measurement system: a review and framework for implementation. Int J Prod Res 58(1):65–86. https://doi.org/10.1080/00207543.2019.1630770
20. Ayyildiz E, Taskin Gumus A (2021) Interval-valued Pythagorean fuzzy AHP method-based supply chain performance evaluation by a new extension of SCOR model: SCOR 4.0. Complex Intell Syst 7(1):559–576. https://doi.org/10.1007/s40747-020-00221-9
21. Lima-Junior FR, Carpinetti LCR (2020) An adaptive network-based fuzzy inference system to supply chain performance evaluation based on SCOR® metrics. Comput Ind Eng 139. https://doi.org/10.1016/j.cie.2019.106191
22. Yadav S, Garg D, Luthra S (2020) Development of IoT based data-driven agriculture supply chain performance measurement framework. J Enterp Inf Manag 34(1):292–327. https://doi.org/10.1108/JEIM-11-2019-0369
23. Beamon BM (1999) Measuring supply chain performance. Int J Oper Prod Manag 19(3):275–292. https://doi.org/10.1108/01443579910249714
24. Ukko J, Saunila M, Rantala T (2020) Connecting relational mechanisms to performance measurement in a digital service supply chain. Prod Planning Control 31(2–3):233–244. https://doi.org/10.1080/09537287.2019.1631466
25. Bechtsis D, Tsolakis N, Vlachos D, Srai JS (2018) Intelligent Autonomous Vehicles in digital supply chains: a framework for integrating innovations towards sustainable value networks. J Clean Prod 181:60–71. https://doi.org/10.1016/j.jclepro.2018.01.173
26. Arzu Akyuz G, Erman Erkan T (2010) Supply chain performance measurement: a literature review. Int J Prod Res 48(17):5137–5155. https://doi.org/10.1080/00207540903089536

Challenges for the Adoption of Industry 4.0 in the Sustainable Manufacturing Supply Chain

Hakeem Owolabi, Luicija Juryte, and Lukman Akanbi

Abstract This book chapter explores the challenges associated with adopting Industry 4.0 technologies in the context of achieving a sustainable manufacturing supply chain. The chapter highlights both general and technology-specific hurdles that organizations encounter when implementing Industry 4.0, such as dealing with data accumulation and compatibility issues with legacy systems, data management complexities, data protection, privacy and cyber attack risks, cost considerations, and workforce upskilling and transition. The chapter emphasizes the importance of addressing these challenges to enable the effective incorporation of Industry 4.0 technologies for sustainability goals. It provides insights and recommendations for mitigating these challenges, including prioritizing sustainability considerations during technology selection and implementation, emphasizing energy efficiency and environmental impact assessments in technology design and deployment, incorporating ethical frameworks and guidelines for data usage, privacy, and fairness in AI and IoT systems, encouraging collaboration among stakeholders to develop industry standards and best practices for sustainable technology adoption, among a few others. By proactively addressing these challenges, organizations can leverage the transformative potential of Industry 4.0 while driving sustainability in their manufacturing supply chains.

Keywords Industry 4.0 · Sustainable manufacturing · Supply chains · Emerging technologies · Waste reduction

L. Juryte · L. Akanbi
Big Data Enteprise and Artificial Intelligence Laboratory (Big-DEAL), Bristol Business School, University of the West of England, Bristol, UK
e-mail: h.owolabi@bolton.ac.uk

H. Owolabi
Greater Manchester Business School (GMBS), University of Bolton, BL3 5AB, Bristol, UK

K E K Vimal et al. (eds.), *Industry 4.0 Technologies: Sustainable Manufacturing Supply Chains*, Environmental Footprints and Eco-design of Products and Processes,
https://doi.org/10.1007/978-981-99-4819-2_12

1 Industry 4.0 and Sustainable Manufacturing Supply Chains

Industry 4.0 (I4.0) has been acknowledged as an excellent opportunity to aggressively push for environmental sustainability within the global manufacturing industry [19, 52]. In an extensive inter-country analysis of manufacturing innovation, [56] described Industry 4.0 (I4.0) as an advanced digitized system that integrates operations, information and communication, state-of-the-art technologies, cyber and physical aspects, people, and the environment over whole value chain. According to [52], industry 4.0 centers on increasing productivity, efficiency and automation of operations through waste reduction, prolonged product life, and lean production. Industry 4.0 enhances traceability and transparency within the production processes, thus facilitating the reduce, reuse, and recycling of end-of-life materials—a key underlying model of the Circular Economy [7]. Through harnessing frontier technologies such as Artificial Intelligence (AI), Blockchain, Big Data Analytics (BDA), Robotics and Automation, 3D printing, and Internet of Things (IoT; industry 4.0 is rapidly transforming manufacturing and its values chains into smarter and more agile production systems, while becoming a catalyst for a more sustainable manufacturing sector of the future [29, 30, 32].

At the heart of sustainable manufacturing is an integrated, efficient, and sustainable supply chain. Seuring and Muller ([57], pp. 1700) defined a sustainable supply chain as "*involving the management of material, information and capital flows as well as cooperation among companies in the supply chain while taking goals from all three dimensions of sustainable development (environmental, social and economic) into account which are derived from customer and stakeholder requirements.*" Supply chains perform central functions in terms of incorporating the demand and supply side of a firm, and have significant economic, social, and environmental impact on societies. For instance, supply chain activities account for up to 17.2% of the US Gross Domestic Product (GDP), up to 10% of the UK's, and 14% of the EU's GDP (EU-ALICE, 2016; US International Trade Administration, 2017; Logistics UK, 2021). Given its sheer size and significance, a slight reduction in supply chain activities will definitely have major environmental sustainability impacts on societies, especially in terms of waste and emissions reduction, decrease in fuel usage, water, and other natural resources [12, 15, 17].

According to [20], one of the key dynamics driving the argument for a greener, more sustainable manufacturing supply chain is the breathtaking speed of globalization and the rise of global markets and production. With the increasing pace of globalization, supply chain activities now cut across national borders, as companies locate production facilities abroad (to access cheaper labor), distribute across borders, and even source materials from foreign suppliers [2–6]. Such increase in supply chain activities [i.e., transportation, procurement, packaging, storage, etc.] at local, national, and global scales ultimately increases carbon emissions and impacts gravely on the environment. In a hugely damning report on the negative impact of supply chain activities on climate change and the environment, the Green House

Gas Protocol report of (2020), put supply chain's contribution to global emission at more than 80% to 90% of emitted greenhouse gasses. To reduce manufacturing's impact on the environment and make the industry more socially and economically sustainable, experts thus believe harnessing industry 4.0 and associated technologies as well as principles offers the strategic path to the future.

2 Emerging Technologies Powering the Industry 4.0 Revolution in Manufacturing

Industry 4.0 combines digital technologies such as Artificial Intelligence (AI), Robotics, Internet of Things (IoT), 3D Printing, Drone Technology, Blockchain, Augmented Reality/Virtual Reality, and Cloud Technology to create new levels of efficiency in manufacturing supply chains [16]. In this way, Industry 4.0 can help enhance sustainability in manufacturing supply chains by improving productivity and reducing energy consumption while optimizing resources for improved environmental performance [16, 25, 46]. According to [25], one of the main motivations for Industry 4.0 is its ability to reduce waste within production processes through automation and data-driven decision making which results in more efficient use of materials and resources throughout the entire supply chain process. Right from raw material extraction all the way through product delivery to customers' doorsteps, Industry 4.0 technologies such as AI, Big Data, IoT, drone technologies, etc., can ensure minimal human intervention and reduced error rates as automated systems replace manual labor tasks where possible [24, 25]. From [16] perspective, these new tech solutions will help minimize pollution caused by inefficient production methods or overconsumption, since fewer resources are needed for each unit produced. By improving production processes via leveraging frontier technologies, Industry 4.0 solutions can reduce emissions that would otherwise be released into our atmosphere from burning fossil fuels used during traditional factory operations [24, 25, 28] while ensuring less packaging materials are being sent offsite for disposal after use [16, 25]. Please see Table 1 for the emerging technologies powering the Industry 4.0 revolution.

Additionally, Industry 4.0-driven solutions like intelligent tracking systems enabled via IoT devices can monitor products at every stage throughout their lifecycle. This thus affords manufacturers greater visibility into their operations enabling them to better manage inventory levels, optimize transportation routes based on real-time information about customer demand patterns, identify areas where further improvements could be made regarding resource utilization and wastage reduction strategies, etc. All these measures taken together result not only in cost savings but also improved sustainability across multiple fronts including reduction/elimination of toxic chemicals used during processing stages, plus increased recycling opportunities when combined with 3D printing technology, thus allowing users access to previously unattainable customization capabilities, with much less requirements for large

Table 1 Emerging technologies powering the Industry 4.0 revolution in manufacturing

Classif	Technology	Definition	Source
Cyber-physical system	**Cyber-Physical Systems (CPS)**	Integration of machines, networks and physical objects, Full automation Interconnectedness Machine to machine communication	[4, 9, 25, 26]
	Internet of Things (IoT) Sensors	Devices that support internet connection and can capture, send and receive information. Applied for object tagging, maintenance-predictability, tracking, Lifecycle monitoring	[28, 47, 55]
	Additive manufacturing 3D printing 3D laser scanning Mass customisation	Producing 3D product layer by layer	[13, 16, 22, 28]
	Advanced Robotics Mobile Robotic Units Cobotics or Collaborative robotics Autonomous Vehicles Drones	Machines that replicate human actions and can work alongside humans (Cobotics), entirely or partly autonomous	[13, 16, 22, 24]
Human-cantered computing	**Immersive Technologies** Augmented Reality (AR) Virtual Reality (VR) Mixed Reality (MR) Holographic Display	Visualising data in an immersed environment	[4, 9, 25, 26, 62]
	Mobile technology Mobile devices Mobile applications	Wireless devices and mobile applications	[9, 16, 22]

(continued)

Table 1 (continued)

Classif	Technology	Definition	Source
Network Technologies	**Cloud computing** Web service Technology Collaboration technology Connected sites	Any services accessed through a cloud provider	[13, 22, 36]
	Blockchain Smart Contracts Supply chain transparency	Distributed ledger technology that provides absolute data transparency	[23, 58]
Computing methodologies	**Simulation and modelling** Digital twins Building Information Modelling (BIM)	Computer imitation of the real-world, testing of various scenarios	[16, 22, 55]
	Artificial intelligence Machine learning Deep learning Natural language processing (NLP) Computer vision	Creation of intelligence that acts and thinks like humans	[13, 16, 22]
Information System	**Big Data** Engineering Analytics	Working with broad volume data, statistics and analytics, storage and processing	[13, 14, 16, 22]
Hardware	**Nanotechnology** Nanomaterials	Controlling atoms or molecules for macroscale fabrication. Nanomaterials can have self-healing properties, increased strength, durability, longer life	[9, 16, 22, 24, 45]

amounts raw material inputs. This Industry 4.0-driven innovations are drastically cutting down on both energy usage and CO_2 output associated with mass-produced items typically found within most traditional retail establishments today.

However, while harnessing Industry 4.0 technologies offers the prospect of enhancing efficiency, productivity, and flexibility, they also bring certain challenges that need to be addressed to ensure sustainability. Experts believe Industry 4.0 presents a number of challenges to sustianble manufacturing supply chains ranging from lack of resources needed to support sustainable supply chain practices [5], lack of skilled personnel who understand how to make the best use of available Industry 4.0 technological solutions efficiently, affordably, and without compromising quality standards set out by industry regulators [2, 3], security concerns as well as vulnerability to cyber attacks from malicious actors. A critical evaluation of a number of

Industry 4.0 technologies and the challenges they present in the context of sustainable manufacturing supply chains have been analyzed below.

3 Challenges of Industry 4.0 Technologies in Sustainable Manufacturing Supply Chains

Despite the probable sustainability and performance gains accruable from harnessing Industry 4.0 technologies, the anticipated benefits to manufacturing supply chains may be hampered in contexts where impediments exceed the enabling factors. According to Mourtzis [48], many Industry 4.0 digital tools [AI, IoT, Blockchain, Big Data, and immersive technologies] being leveraged by supply chains are novel and therefore, constantly undergoing large-scale updates. This potentially makes them vulnerable to diverse operational risks [security, safety, and privacy] [41, 42, 63], in addition to other implementation challenges [i.e., high development and operational cost, skill shortage, lack of available quality data, etc.] [27, 28, 43]. Furthermore, the global supply chain industry is dominated by Small and Medium-Scale Enterprises (SMEs) [62, 37], despite the presence of other very large global supply chains. This dominance of SMEs in the industry could possibly impair the adoption rate for Industry 4.0 technologies and impact supply chain performance to deliver anticipated outcomes.

Similarly, despite the increasing momentum for the adoption of digital supply chains within the global manufacturing industry [35, 37, 49], there are growing tales of low levels of integration across many supply chain networks, especially in the UK. Recent surveys of the UK manufacturing industry indicated fewer UK firms have installed control tower-based real-time monitoring of their supply chains. Beyond the adoption and integration of Industry 4.0 technologies in supply chains, there is also the challenge of performance measurement in digitalized supply chain settings [15, 32], Association of Supply Chain Management 2022). As compared with the traditional or non-digital supply chains (non-DSCs)—which have considerably more developed performance measures and measurement systems—it remains unclear whether supply chains driven by Industry 4.0 technologies will have to evolve or adapt traditional supply chain performance measures into digitally driven ones [35, 37, 49].

Additionally, in terms of performance measurement, advanced technology-driven supply chains offer different operational realities that could render performance indicators of traditional supply chains inadequate, in a digital supply chain setting [1, 12, 18]. This potential irregularity in performance measures and measurement has been recognized in the supply chain management (SCM) domain, given the concurrent development and recent release of the digital version of the SCOR model known as the SCOR Digital Standard (SCOR DS) (Association for Supply Chain Management [8]). This became crucial as the current SCOR model fails to adequately reflect the operational characteristics of digital supply chains, thus necessitating the need to update the SCOR model with the relevant body of digital knowledge and essential sustainability and resilience standards (Mizell-Pleasant 2022), [32].

Whilst the advantages of Industry 4.0 have been widely acknowledged, experts have also pointed out other difficulties that might accompany the digitalization of supply chains. For example [40], contended that by opting to digitalize supply chains, the data generation becomes much easy and cheap. As a result, more data output is accumulated, which can further increase supply chain complexities and result in higher costs of managing, storing, and retrieving data [61] also highlighted some challenges such as the issues relating to trust and security which arises with the use of digitalized supply chains. In addition, since interconnectivity across networks is central to digilized supply chains, this therefore increases the possibility of cyber threats and attacks on networks and could potentially spread across the entire supply chain, triggering system-wide disruptions [33]. As such, vital digital supply chain enabling technologies, i.e., smart systems and IoT sensors, have been noted to be extremely vulnerable to cyberware attacks and data leakage [10].

Though emerging technologies are at the core of Industry 4.0 due to the pervasiveness and speed of progression, there are many uncertainties concerning the application in practice and advancement trajectories [16]. Studies like [16, 21, 46] have highlighted the vast amount of different heterogeneous technologies currently being used and researched, that could be included in Industry 4.0 strategy. In addition, the technological landscape is continually changing, with new technologies emerging in various areas [25], thus making restructuring of manufacturing operations a constant challenge. Given the huge interest in this topic, several other authors have taken the initiative to summarize the most significant barriers discussed in the literature [11, 50, 51, 56]. Table 2 for Challenges to Industry 4.0 Implementation in Sustainable Manufacturing Supply Chains.

4 Technology-Level Challenges of Industry 4.0 Adoption for Sustainable Manufacturing Supply Chain Management

Adopting Industry 4.0 technologies to improve sustainability in manufacturing supply chains can also pose several technology-level challenges. While these technologies offer great potential, addressing the associated challenges highlighted below is essential for successful implementation:

1. **Internet of Things (IoT)**:

    a. **Data privacy and security**: The huge amount of data shunned by IoT devices can raise concerns regarding data privacy and security. Protecting sensitive information and ensuring safe data transmission and storage is vital for supply chains to maintain sustainability and trust.

Table 2 Challenges to Industry 4.0 Implementation in Sustainable Manufacturing Supply Chains

Barriers	Definition	Source
Cost of the implementation	The cost of technology is high, technology is not fully developed and continually advancing, training to use technologies is also high	[11, 39, 50, 55]
Technology non-acceptance	Supply chain industry is dominated by low-tier SMEs with meager technology and administrative capacity to leverage advanced technologies	[31, 44, 51, 55]
Poor Implementation in the Industry	Technology adoption is still quite low in many manufacturing supply chains across the globe	[24, 55]
Dependence on complex value chain	Manufacturing industry is dependent on downstream and upstream supply chain	[24, 53, 55]
Low technology maturity level and a wide variety	The technological landscape is ever-changing, new technologies appearing continuously	[53, 56]
The high complexity of the construction project	Each project is unique and highly customizable	[51, 55]
High requirements	Maintaining and utilizing technology needs specialist	[34, 55]
Lack of knowledge	I4.0 research in the Industry does not guide organizations on how to start and continue with revolution	[38, 50, 51, 56, 60]
Lack of skilled employees	There is also a lack of skilled employees who would be able to use such technologies	[53, 55, 56, 59]
Lack of standards	Technological advancement is not regulated, safeguarding customers' interest is challenging	[11, 56]
Risk of security breaches	Cybersecurity concerns and fair of losing data	[11, 51, 56]
Poor long-term planning	Long-term outcomes and risk assessments are not always considered for the long-term well-being of the construction project	[54]
Insufficient support	Support for technology users and organizations implementing it	[11, 51]

b. **Electronic waste**: IoT devices, if not properly managed, can contribute to electronic waste generation. The disposal and recycling of these devices need to be handled responsibly to mitigate environmental impact. At the moment, it is unclear whether many supply chains have practices or strategies in place to address these challenges.

2. **Big Data and Analytics**:

    a. **Energy consumption**: The processing and storage of large volumes of data require substantial computing power and energy consumption. To avoid increasing their carbon footprints as they leverage technologies, supply chains require efficient data center management and energy-saving measures to minimize the environmental footprint.

Ethical Data Usage:

b. Handling a large amount of third-party datasets can present huge challenges to supply chains. Hence, ethical collection, storage, and use of data are essential. Maintaining transparency and ensuring compliance with privacy regulations are also critical to upholding sustainability principles in supply chains.

3. **Artificial Intelligence (AI) and Machine Learning (ML)**:

    a. **Energy consumption**: AI and ML algorithms often need substantial computational power, thus resulting in more energy consumption. For a supply chain hoping to go green, optimizing algorithms and utilizing energy-efficient hardware remains the most viable way to help address this challenge.
    b. **Bias and fairness**: AI systems can perpetuate biases if not carefully designed and trained. Ensuring fairness and eliminating biases in decision-making processes is important for sustainable and equitable outcomes.

Ethical Implications:

c. As AI systems make autonomous decisions, ethical considerations arise, with many supply chains lacking the required funds to invest in the needed expertise to develop ethically compliant solutions. However, ensuring responsible AI use, transparency, and accountability are vital for sustainable manufacturing practices.

4. **Robotics and Automation**:

    a. **Workforce displacement**: The adoption of robotics and automation technologies may result in the displacement of jobs or changes in the workforce. As such, ensuring a just transition for affected workers and providing retraining opportunities are crucial for social sustainability in supply chains.
    b. **Environmental impact**: While automation can improve efficiency, it may also lead to increased resource consumption during the production and disposal of robotic systems. Managing the life cycle of automation technologies and considering their environmental impact is necessary for sustainability.

5. **Additive Manufacturing (3D Printing)**:

    a. **Material usage and waste**: Although 3D printing can reduce material waste compared to traditional manufacturing, the sustainability of additive manufacturing depends on the responsible sourcing and disposal of materials used in the process. Supply chains therefore need to take cognizance of the end-to-end lifecycle management of addictive manufacturing raw materials while investing in sustainable ones.

b. **Energy consumption**: 3D printing can consume significant energy, especially for larger scale production. Employing energy-efficient 3D printers and optimizing printing processes can help mitigate this challenge, thus ensuring a more sustainable production and procurement process.

6. **Blockchain**:

   a. **Energy consumption**: The energy requirements for blockchain technology can be significant. Hence, while there is a general call for the adoption of blockchains, especially in the aspect of tracking carbon emission footprints of manufacturing supply chains, there is a vital need to consider employing energy-efficient consensus mechanisms and exploring alternative energy sources to help reduce its environmental impact.
   b. **Supply chain transparency**: Although blockchain can enhance supply chain transparency and traceability, its implementation requires careful consideration of data accuracy, data management, and the incorpration of various procedures for smooth exchange of information.

Based on the above, addressing the challenges posed by Industry 4.0 technologies for achieving a sustainable manufacturing supply chain requires therefore proactive measures. Experts have suggested a number of strategies that include:

1. Prioritizing sustainability considerations during technology selection and implementation.
2. Emphasizing energy efficiency and environmental impact assessments in technology design and deployment.
3. Incorporating ethical frameworks and guidelines for data usage, privacy, and fairness in AI and IoT systems.
4. Encouraging collaboration among stakeholders to develop industry standards and best practices for sustainable technology adoption.
5. Investing in workforce development and reskilling programs to mitigate the social impacts of automation and ensure a just transition.

5 Conclusion and Implication

In conclusion, this book chapter has shed light on the challenges organizations face when adopting Industry 4.0 technologies to achieve a sustainable manufacturing supply chain. The chapter has highlighted both general challenges, such as data accumulation and compatibility issues, as well as technology-specific hurdles like data management complexities, cybersecurity risks, and the need for workforce upskilling. It has emphasized the criticality of addressing these challenges to successfully integrate Industry 4.0 technologies for sustainability goals. The chapter has provided valuable insights and recommendations for mitigating these challenges, including prioritizing sustainability considerations, incorporating ethical frameworks, and fostering collaboration among stakeholders. By proactively addressing these challenges, organizations can harness the transformative potential of Industry 4.0 while driving sustainability in their manufacturing supply chains.

The implications of this chapter are significant for practitioners, researchers, and policymakers involved in the adoption and implementation of Industry 4.0 technologies. Understanding the challenges discussed enables organizations to make informed decisions and develop effective strategies to overcome hurdles. The chapter's insights highlight the need to consider sustainability as a key criterion during technology selection and implementation, ensuring that energy efficiency and environmental impact assessments are integral parts of design and deployment processes. The incorporation of ethical frameworks and guidelines for data usage, privacy, and fairness is crucial to maintain trust and promote responsible AI and IoT systems. Furthermore, collaboration among stakeholders to develop industry standards and best practices fosters a collective effort towards sustainable technology adoption. By heeding these implications, organizations can navigate the challenges of Industry 4.0 adoption and unlock its potential to drive sustainable manufacturing supply chains.

References

1. Agami N, Saleh M, Rasmy M (2012) (2012) Supply chain performance measurement approaches: review and classification. J Organ Manag Studies 2012:1–20
2. Agrawal P, Narain R (2018) Digital supply chain management: an overview. Mater Sci Eng 455(1):1–15
3. Akbari M, Hopkins JL (2022) Digital technologies as enablers of supply chain sustainability in an emerging economy. To be published in *Operations Management Research* [preprint]. Available from: https://doi.org/10.1007/s12063-021-00226-8. Accessed 4 August 2022
4. Alcácer V, Cruz-Machado V (2019) Scanning the industry 4.0: A literature review on technologies for manufacturing systems. Eng Sci Techn Intern J 22 (3):899–919. https://doi.org/10.1016/j.jestch.2019.01.006
5. Alexander KL, Lopes B, Ricchetti-Masterson K, Yeatts BK (2015) *'Cross-sectional studies'*. Eric Notebook 2nd Ed. North Carolina: UNC Gillings School of Global Public Health [Online]. Available at: https://sph.unc.edu/wp-content/uploads/sites/112/2015/07/nciph_ERIC8.pdf. Accessed 2 September 2022
6. Alias Z, Zawawi EMA, Yusof K, Aris NM (2014) Determining critical success factors of project management practice: a conceptual framework. Procedia Soc Behav Sci 153(1):61–69
7. Antikainen M, Uusitalo T, Kivikytö-Reponen P (2018) Digitalisation as an enabler of circular economy. In: Procedia CIRP. 1 January 2018 (no place) Elsevier B.V. pp. 45–49. https://doi.org/10.1016/j.procir.2018.04.027
8. Association for Supply Chain Management (2022) The SCOR model explained: announcing the new SCOR digital standard (SCOR DS) [blog]. Available from: https://www.ascm.org/corporate-transformation/standards-tools/scor-ds/. Accessed 7 September 2022
9. Bai C, Dallasega P, Orzes G, Sarkis J (2020) Industry 4.0 technologies assessment: A sustainability perspective. Int J Prod Econom 229:107776. https://doi.org/10.1016/j.ijpe.2020.107776
10. Balasundaram J (2021) A novel optimized bat extreme learning intrusion detection system for smart internet of things networks. Int J Commun Syst 34(7):4729–4730

11. Benitez GB, Ayala NF, Frank AG (2020) Industry 4.0 innovation ecosystems: an evolutionary perspective on value cocreation. Int J Production Econ 228. https://doi.org/10.1016/j.ijpe.2020.107735
12. Bhattacharya A, Mohapatra P, Kumar V, Dey PK, Brady M, Tiwari MK, Nudurupati SS (2014) Green supply chain performance measurement using fuzzy ANP-based balanced scorecard: a collaborative decision-making approach. Prod Planning Control 25(8):698–714
13. Bigliardi B, Bottani E, Casella G (2020) Enabling technologies, application areas and impact of industry 4.0: A bibliographic analysis. Procedia Manuf 42:322–326 (2019). https://doi.org/10.1016/j.promfg.2020.02.086
14. Bilal M, Oyedele LO, Qadir J, Munir K, Ajayi SO, Akinade OO, Owolabi, HA, Alaka HA, Pasha M (2016) Big data in the construction industry: A review of present status, opportunities, and future trends. Adv Eng Informatics 30(3):500–521. https://doi.org/10.1016/j.aei.2016.07.001
15. Bolstorff P, Rosenbaum R (2012) Supply chain excellence: a handbook for dramatic improvement using the SCOR model, 3rd edn. American Management Association, New York
16. Bonaccorsi A, Chiarello F, Fantoni G, Kammering H (2020) Emerging technologies and industrial leadership. A Wikipedia-based strategic analysis of Industry 4.0. Expert Syst Appl 160:113645. https://doi.org/10.1016/j.eswa.2020.113645
17. Bonache J, Festing M (2020) Research paradigms in international human resource management: An epistemological systematisation of the field. German J Human Resour Manag 34(2):99–123
18. Cai J, Liu X, Xiao Z, Liu J (2009) Improving supply chain performance management: a systematic approach to analyzing iterative KPI accomplishment. Decis Support Syst 46(2):512–521
19. Charter M (2018) Designing for the circular economy. 1st edition. (no place) Routledge
20. Chau KY, Tang YM, Liu X, Ip YK, Tao Y (2021) Investigation of critical success factors for improving supply chain quality management in manufacturing. Enterp Inf Syst 15(10):1418–1437
21. Chiarello F, Trivelli L, Bonaccorsi A, Fantoni G (2018) Extracting and mapping industry 4.0 technologies using Wikipedia. Comput Indus 100:244–257. https://doi.org/10.1016/j.compind.2018.04.006
22. Chiarello F, Trivelli L, Bonaccorsi A, Fantoni G (2018) Extracting and mapping industry 4.0 technologies using Wikipedia. Comp Industry. 100:244–257 (September 2017). https://doi.org/10.1016/j.compind.2018.04.006
23. Cong LW, He Z (2019) Blockchain disruption and smart contracts. Rev Financ Stud 32(5):1754–1797. https://doi.org/10.1093/rfs/hhz007
24. Craveiro F, Duarte JP, Bartolo H, Bartolo PJ (2019) Additive manufacturing as an enabling technology for digital construction: a perspective on construction 4.0. Autom Constr 103:251–267. https://doi.org/10.1016/j.autcon.2019.03.011
25. Culot G, Nassimbeni G, Orzes G, Sartor M (2020) Behind the definition of Industry 4.0: analysis and open questions. Int J Prod Econ 226:107617. https://doi.org/10.1016/j.ijpe.2020.107617
26. Da Silva A, Almeida I (2020) Towards INDUSTRY 4.0 | a case STUDY in ornamental stone sector. Resour Policy 67:101672. https://doi.org/10.1016/j.resourpol.2020.101672
27. Dallasega P, Rauch E, Linder C (2018) Industry 4.0 as an enabler of proximity for construction supply chains: a systematic literature review. Comput Indus 99:205–225. https://doi.org/10.1016/j.compind.2018.03.039
28. Dalmarco G, Ramalho FR, Barros AC, Soares AL (2019) Providing industry 4.0 technologies: the case of a production technology cluster. J High Technol Manag Res 30(2):100355. https://doi.org/10.1016/j.hitech.2019.100355
29. Dweekat AJ, Hwang G, Park J (2017) A supply chain performance measurement approach using the internet of things: toward more practical SCPMS. Ind Manag Data Syst 117(2):267–286
30. Dwivedi A, Paul SK (2022) A framework for digital supply chains in the era of circular economy: implications on environmental sustainability. Bus Strateg Environ 31:1249–1274

31. Edwards DJ, Pärn E, Love PED, El-Gohary H (2017) Research note: Machinery, manumission, and economic machinations. J Bus Res 70:391–394. https://doi.org/10.1016/j.jbusres.2016.08.012
32. Es-Satty A, Lemghari R, Okar C (2020) Supply chain digitalization overview SCOR model implication. In: 2020 IEEE 13th international colloquium of logistics and supply chain management (LOGISTIQUA) New Jersey: Institute of Electrical and Electronics Engineers, pp 1–7
33. Fadi AT, Deebak BD (2020) Seamless authentication: for IoT-big data technologies in smart industrial application systems. IEEE Trans Industr Inf 17(4):2919–2927
34. Fisher LH, Edwards DJ, Pärn EA, Aigbavboa CO (2018) Building design for people with dementia: a case study of a UK care home. Facilities 36(7–8):349–368. https://doi.org/10.1108/F-06-2017-0062
35. Frank AG, Dalenogare LS, Ayala NF (2019) Industry 4.0 technologies: implementation patterns in manufacturing companies. Int J Prod Econ 210:15–26. https://doi.org/10.1016/j.ijpe.2019.01.004
36. Frank AG, Dalenogare LS, Ayala NF (2019) Industry 4.0 technologies: Implementation patterns in manufacturing companies. Int J Prod Econom 210:15–26. https://doi.org/10.1016/j.ijpe.2019.01.004
37. Garay-Rondero CL, Martinez-Flores JL, Smith NR, Caballero Morales SO, Aldrette-Malacara A (2019) Digital supply chain model in Industry 4.0. J Manuf Technol Manag. https://doi.org/10.1108/JMTM-08-2018-0280
38. Glas AH, Kleemann FC (2016) The Impact of Industry 4.0 on procurement and supply management : a conceptual and qualitative analysis. Int J Bus Manag Invention 5(6):55–66
39. Griffin A, Hughes R, Freeman C, Illingworth J, Hodgson T, Lewis M, Perez E (2019) Using advanced manufacturing technology for smarter construction. Proc Inst Civ Eng Civ Eng 172(6):15–21. https://doi.org/10.1680/jcien.18.00051
40. Hazen BT, Boone CA, Ezell JD, Jones-Farmer LA (2014) Data quality for data science, predictive analytics, and big data in supply chain management: an introduction to the problem and suggestions for research and applications. Int J Prod Econ 154(1):72–80
41. Islam A (2018) performance evaluation of supply chain network in apparel industry: a case study. Masters Thesis, Bangladesh University of Engineering and Technology
42. Jamehshooran GB, Shaharoun M, Norehan Haron H (2015) Assessing supply chain performance through applying the SCOR model. Inte J Supply Chain Manag 4(1):1–11
43. Kamble SS, Gunasekaran A, Gawankar SA (2018) Sustainable Industry 4.0 framework: a systematic literature review identifying the current trends and future perspectives. Process Saf Environ Prot 117:408–425. https://doi.org/10.1016/j.psep.2018.05.009
44. Karadayi-Usta S (2020) An interpretive structural analysis for industry 4.0 adoption challenges. IEEE Trans Eng Manag 67(3):973–978. https://doi.org/10.1109/TEM.2018.2890443
45. Khandve P (2014) Nanotechnology for building material. Int J Basic Appli Res 4:146–151
46. Klingenberg CO, Borges MAV, Antunes JAV (2019) Industry 4.0 as a data-driven paradigm: a systematic literature review on technologies. J Manuf Technol Manag (88881). https://doi.org/10.1108/JMTM-09-2018-0325
47. Maskuriy R, Selamat A, Ali KN, Maresova P, Krejcar O (2019) Industry 4.0 for the construction industry—how ready Is the industry? Applied Sci 9(14):2819. https://doi.org/10.3390/app9142819
48. Mourtzis D (2020) Adaptive scheduling in the era of cloud manufacturing. In: Sokolov B, Ivanov D, Dolgui A (eds) Scheduling in industry 4.0 and cloud manufacturing. New York: Springer, pp 61–85
49. Mubarak MF, Petraite M (2020) Industry 4.0 technologies, digital trust and technological orientation: what matters in open innovation? Technol Forecasting Soc Change 161:120332. https://doi.org/10.1016/j.techfore.2020.120332
50. Newman C, Edwards D, Martek I, Lai J, Thwala WD, Rillie I (2020) Industry 4.0 deployment in the construction industry: a bibliometric literature review and UK-based case study. Smart Sustain Built Environ. https://doi.org/10.1108/SASBE-02-2020-0016

51. Oesterreich TD, Teuteberg F (2016) Understanding the implications of digitisation and automation in the context of Industry 4.0: a triangulation approach and elements of a research agenda for the construction industry. Comput Indus 83:121–139. https://doi.org/10.1016/j.compind.2016.09.006
52. Okorie O, Salonitis K, Charnley F, Moreno M, Turner C, Tiwari A (2018) Digitisation and the circular economy: a review of current research and future trends. Energies 11(11):3009. https://doi.org/10.3390/en11113009
53. Ortt R, Stolwijk C (2020) Implementing industry 4. 0 : assessing the current state. J Manuf Technol Manag. https://doi.org/10.1108/JMTM-07-2020-0284
54. Pereira T, Barreto L, Amaral A (2017) Network and information security challenges within Industry 4.0 paradigm. Procedia Manuf 13:1253–1260. https://doi.org/10.1016/j.promfg.2017.09.047
55. Perrier N, Bled A, Bourgault M, Cousin N, Danjou C, Pellerin R, Roland T (2020) Construction 4.0: a survey of research trends. J Inf Technol Construction 25:416–437. https://doi.org/10.36680/J.ITCON.2020.024
56. Raj A, Dwivedi G, Sharma A, Lopes de Sousa Jabbour AB, Rajak S (2020) Barriers to the adoption of industry 4.0 technologies in the manufacturing sector: an inter-country comparative perspective. Int J Prod Econ 224:107546. https://doi.org/10.1016/j.ijpe.2019.107546
57. Seuring S, Müller M (2008) From a literature review to a conceptual framework for sustainable supply chain management. J Clean Prod 16(15):1699–1710
58. Sunny J, Undralla N, Madhusudanan Pillai V (2020) Supply chain transparency through blockchain-based traceability: An overview with demonstration. Comp Ind Eng 150:106895. https://doi.org/10.1016/j.cie.2020.106895
59. Trotta D, Garengo P (2018) Industry 4.0 key research topics: a bibliometric review. 2018 7th international conference on industrial technology and management, ICITM 2018. 2018-Janua pp 113–117. https://doi.org/10.1109/ICITM.2018.8333930
60. Zabidin NS, Belayutham S, Ibrahim CKIC (2020) A bibliometric and scientometric mapping of Industry 4.0 in construction. J Inf Technol Construction 25:287–307. https://doi.org/10.36680/j.itcon.2020.017
61. Zhang H, Nakamura T, Sakurai K (2019) Security and trust issues on digital supply chain. In: Proceedings—IEEE 17th international conference on dependable, autonomic and secure computing, IEEE 17th international conference on pervasive intelligence and computing, IEEE 5th international conference on cloud and big data computing, 4th Cyber Science
62. de Paula Ferreira W, Armellini F, De Santa-Eulalia LA, William (2020) Simulation in industry 4.0: a state-of-the-art review. Comput Indus Eng 149:106868. https://doi.org/10.1016/j.cie.2020.106868
63. Özkanlısoy Ö, Akkartal E (2021) Digital transformation in supply chains: Current applications, contributions and challenges. Bus Manag Studies Int J 9(1):32–55

MIX
Papier aus verantwortungsvollen Quellen
Paper from responsible sources
FSC® C105338

If you have any concerns about our products,
you can contact us on
ProductSafety@springernature.com

In case Publisher is established outside the EU,
the EU authorized representative is:
Springer Nature Customer Service Center GmbH
Europaplatz 3, 69115 Heidelberg, Germany

Printed by Libri Plureos GmbH
in Hamburg, Germany